I0868184

IMPROVING GENETICS EDUCATION IN GRADUATE AND CONTINUING HEALTH PROFESSIONAL EDUCATION

WORKSHOP SUMMARY

Adam C. Berger, Samuel G. Johnson, Sarah H. Beachy,
and Steve Olson, *Rapporteurs*

Roundtable on Translating Genomic-Based
Research for Health

Board on Health Sciences Policy

INSTITUTE OF MEDICINE
OF THE NATIONAL ACADEMIES

THE NATIONAL ACADEMIES PRESS
Washington, D.C.
www.nap.edu

THE NATIONAL ACADEMIES PRESS 500 Fifth Street, NW Washington, DC 20001

NOTICE: The workshop that is the subject of this workshop summary was approved by the Governing Board of the National Research Council, whose members are drawn from the councils of the National Academy of Sciences, the National Academy of Engineering, and the Institute of Medicine.

This project was supported by contracts between the National Academy of Sciences and the American Academy of Nursing (unnumbered contract); American College of Medical Genetics and Genomics (unnumbered contract); American Heart Association (unnumbered contract); American Medical Association (unnumbered contract); American Society of Human Genetics (unnumbered contract); Association for Molecular Pathology (unnumbered contract); Blue Cross and Blue Shield Association (unnumbered contract); College of American Pathologists (unnumbered contract); Department of Veterans Affairs (Contract No. VA248-P-1528 and Contract No. VA240-14-C-0037); Eli Lilly and Company (unnumbered contract); Genetic Alliance (unnumbered contract); Health Resources and Services Administration (Contract No. HHSH250200976014I, Order Nos. HHSH25034017T and HHSH25034021T); International Society for Cardiovascular Translational Research (unnumbered contract); Kaiser Permanente Program Offices Community Benefit II at the East Bay Community Foundation (Contract No. 20121257); Life Technologies (unnumbered contract); Merck & Co., Inc. (Contract No. CMO-140505-000393); National Cancer Institute (Contract No. HHSN263201200074I, TO#5); National Human Genome Research Institute (Contract No. HHSN263201200074I, TO#5); National Institute of Mental Health (Contract No. HHSN263201200074I, TO#5); National Institute of Nursing Research (Contract No. HHSN263201200074I, TO#5); National Institute on Aging (Contract No. HHSN263201200074I, TO#5); National Society of Genetic Counselors (unnumbered contract); Northrop Grumman Health IT (unnumbered contract); Pfizer Inc. (unnumbered contract); and PhRMA (unnumbered contract). The views presented in this publication do not necessarily reflect the views of the organizations or agencies that provided support for the activity.

International Standard Book Number-13: 978-0-309-31605-7
International Standard Book Number-10: 0-309-31605-7

Additional copies of this workshop summary are available for sale from the National Academies Press, 500 Fifth Street, NW, Keck 360, Washington, DC 20001; (800) 624-6242 or (202) 334-3313; http://www.nap.edu.

For more information about the Institute of Medicine, visit the IOM home page at: **www.iom.edu.**

Copyright 2015 by the National Academy of Sciences. All rights reserved.

Printed in the United States of America

Suggested citation: IOM (Institute of Medicine). 2015. *Improving genetics education in graduate and continuing health professional education: Workshop summary.* Washington, DC: The National Academies Press.

"Knowing is not enough; we must apply.
Willing is not enough; we must do."

—Goethe

INSTITUTE OF MEDICINE

OF THE NATIONAL ACADEMIES

Advising the Nation. Improving Health.

THE NATIONAL ACADEMIES

Advisers to the Nation on Science, Engineering, and Medicine

The **National Academy of Sciences** is a private, nonprofit, self-perpetuating society of distinguished scholars engaged in scientific and engineering research, dedicated to the furtherance of science and technology and to their use for the general welfare. Upon the authority of the charter granted to it by the Congress in 1863, the Academy has a mandate that requires it to advise the federal government on scientific and technical matters. Dr. Ralph J. Cicerone is president of the National Academy of Sciences.

The **National Academy of Engineering** was established in 1964, under the charter of the National Academy of Sciences, as a parallel organization of outstanding engineers. It is autonomous in its administration and in the selection of its members, sharing with the National Academy of Sciences the responsibility for advising the federal government. The National Academy of Engineering also sponsors engineering programs aimed at meeting national needs, encourages education and research, and recognizes the superior achievements of engineers. Dr. C. D. Mote, Jr., is president of the National Academy of Engineering.

The **Institute of Medicine** was established in 1970 by the National Academy of Sciences to secure the services of eminent members of appropriate professions in the examination of policy matters pertaining to the health of the public. The Institute acts under the responsibility given to the National Academy of Sciences by its congressional charter to be an adviser to the federal government and, upon its own initiative, to identify issues of medical care, research, and education. Dr. Victor J. Dzau is president of the Institute of Medicine.

The **National Research Council** was organized by the National Academy of Sciences in 1916 to associate the broad community of science and technology with the Academy's purposes of furthering knowledge and advising the federal government. Functioning in accordance with general policies determined by the Academy, the Council has become the principal operating agency of both the National Academy of Sciences and the National Academy of Engineering in providing services to the government, the public, and the scientific and engineering communities. The Council is administered jointly by both Academies and the Institute of Medicine. Dr. Ralph J. Cicerone and Dr. C. D. Mote, Jr., are chair and vice chair, respectively, of the National Research Council.

www.national-academies.org

PLANNING COMMITTEE[1]

BRUCE BLUMBERG (*Co-Chair*), Institutional Director of Graduate Medical Education, Northern California, Kaiser Permanente, The Permanente Medical Group

JOAN A. SCOTT (*Co-Chair*), Chief, Genetics Services Branch, Division of Services for Children with Special Health Needs, Maternal and Child Health Bureau, Health Resources and Services Administration

DAVID A. DAVIS, Senior Director, Continuing Education and Performance Improvement, Association of American Medical Colleges

MICHAEL J. DOUGHERTY, Director of Education, American Society of Human Genetics

W. GREGORY FEERO, Contributing Editor, *Journal of the American Medical Association*

KATHERINE JOHANSEN TABER, Director, Personalized Medicine; Assistant Secretary, Council on Science and Public Health, American Medical Association

JANET K. WILLIAMS, Representative of the American Academy of Nursing; Professor of Nursing, Chair of the Behavioral and Social Science Research Institutional Review Board, University of Iowa

Fellow

SAMUEL G. JOHNSON, American Association of Colleges of Pharmacy, American College of Clinical Pharmacy Anniversary Fellow

IOM Staff

ADAM C. BERGER, Project Director
SARAH H. BEACHY, Associate Program Officer
MEREDITH HACKMANN, Senior Program Assistant

[1]Institute of Medicine planning committees are solely responsible for organizing the workshop, identifying topics, and choosing speakers. The responsibility for the published workshop summary rests with the workshop rapporteurs and the institution.

ROUNDTABLE ON TRANSLATING GENOMIC-BASED RESEARCH FOR HEALTH[1]

GEOFFREY GINSBURG (*Co-Chair*), Director, Center for Genomic Medicine, Institute for Genomic Sciences and Policy, Duke University, Durham, NC
SHARON TERRY (*Co-Chair*), President and Chief Executive Officer, Genetic Alliance, Washington, DC
NAOMI ARONSON, Executive Director, Technology Evaluation Center, Blue Cross and Blue Shield Association, Chicago, IL
EUAN ANGUS ASHLEY, Representative of the American Heart Association; Director, Center for Inherited Cardiovascular Disease, Stanford University School of Medicine, Palo Alto, CA
PAUL R. BILLINGS, former Chief Medical Officer, Life Technologies, Carlsbad, CA
BRUCE BLUMBERG, Institutional Director of Graduate Medical Education, Northern California Kaiser Permanente, The Permanente Medical Group, Oakland, CA
PAMELA BRADLEY, Staff Fellow, Personalized Medicine Staff, Office of In Vitro Diagnostics and Radiological Health, Center for Devices and Radiological Health, U.S. Food and Drug Administration, Silver Spring, MD (until October 2014)
PHILIP J. BROOKS, Health Scientist Administrator, Office of Rare Diseases Research, National Center for Advancing Translational Sciences, National Institutes of Health, Bethesda, MD
ANN CASHION, Scientific Director, National Institute of Nursing Research, National Institutes of Health, Bethesda, MD
C. THOMAS CASKEY, Professor, Baylor College of Medicine, Houston, TX (until June 2014)
ROBERT B. DARNELL, President and Scientific Director, New York Genome Center; Investigator, Howard Hughes Medical Institute, Heilbrunn Cancer Professor and Senior Physician, Head, Laboratory of Molecular Neuro-Oncology, Rockefeller University, New York, NY
MICHAEL J. DOUGHERTY, Director of Education, American Society of Human Genetics, Bethesda, MD

---

[1]Institute of Medicine forums and roundtables do not issue, review, or approve individual documents. The responsibility for the published workshop summary rests with the workshop rapporteurs and the institution.

VICTOR J. DZAU, President and Chief Executive Officer, Duke University Health System; Chancellor for Health Affairs, Duke University, Durham, NC (until June 2014)

W. GREGORY FEERO, Contributing Editor, *Journal of the American Medical Association*, Chicago, IL

ANDREW N. FREEDMAN, Branch Chief, Clinical and Translational Epidemiology Branch, Epidemiology and Genetics Research Program, Division of Cancer Control and Population Sciences, National Cancer Institute, Rockville, MD

JENNIFER L. HALL, Representative of the International Society for Cardiovascular Translational Research; Associate Professor of Medicine, University of Minnesota, Minneapolis

RICHARD J. HODES, Director, National Institute on Aging, Bethesda, MD

MUIN KHOURY, Director, National Office of Public Health Genomics, Centers for Disease Control and Prevention, Atlanta, GA

GABRIELA LAVEZZARI, Assistant Vice President, Scientific Affairs, PhRMA, Washington, DC

THOMAS LEHNER, Director, Office of Genomics Research Coordination, National Institute of Mental Health, Bethesda, MD

DEBRA LEONARD, Representative of the College of American Pathologists; Professor and Chair of Pathology at the University of Vermont College of Medicine; Physician Leader of Pathology and Laboratory Medicine at Fletcher Allen Health Care, University of Vermont College of Medicine, University of Vermont, Burlington

TERI A. MANOLIO, Director, Division of Genomic Medicine, National Human Genome Research Institute, Rockville, MD

OLUFUNMILAYO F. OLOPADE, Walter L. Palmer Distinguished Service Professor of Medicine; Director, Center for Clinical Cancer Genetics; Associate Dean for Global Health, University of Chicago, IL (until April 2014)

MICHELLE A. PENNY, Senior Director, Translational Medicine Group, Eli Lilly and Company, Indianapolis, IN

AIDAN POWER, Vice President and Head, PharmaTx Precision Medicine, Pfizer Inc., Groton, CT (until July 2014)

VICTORIA M. PRATT, Representative of the Association for Molecular Pathology; Associate Professor of Clinical Medical and Molecular Genetics and Director, Pharmacogenomics Diagnostic Laboratory, Department of Medical and Molecular Genetics, Indiana University School of Medicine, Indianapolis

RONALD PRZYGODZKI, Associate Director for Genomic Medicine and Acting Director of Biomedical Laboratory Research and Development, Department of Veterans Affairs, Washington, DC

ALLEN D. ROSES, President and Chief Operating Officer, Cabernet, Shiraz and Zinfandel Pharmaceuticals; and Jefferson–Pilot Professor of Neurobiology and Genetics, Professor of Medicine (Neurology); Director, Deane Drug Discovery Institute; Senior Scholar, Fuqua School of Business, R. David Thomas Executive Training Center, Duke University, Durham, NC (until May 2014)

JOAN A. SCOTT, Chief, Genetic Services Branch, Division of Services for Children with Special Health Needs, Maternal and Child Health Bureau, Rockville, MD

SAM SHEKAR, Chief Medical Officer, Health Information Technology Program, Northrop Grumman Information Systems, McLean, VA

KATHERINE JOHANSEN TABER, Director, Personalized Medicine; Assistant Secretary, Council on Science and Public Health, American Medical Association, Chicago, IL

DAVID VEENSTRA, Professor, Pharmaceutical Outcomes Research and Policy Program, Department of Pharmacy, University of Washington, Seattle

MICHAEL S. WATSON, Executive Director, American College of Medical Genetics and Genomics, Bethesda, MD

DANIEL WATTENDORF, Deputy Chief, Medical Innovations, Department of the Air Force; Program Manager, Defense Advanced Research Projects Agency/Defense Sciences Office, Arlington, VA

CATHERINE A. WICKLUND, Past President, National Society of Genetic Counselors; Director, Graduate Program in Genetic Counseling; Associate Professor, Department of Obstetrics and Gynecology, Northwestern University, Chicago, IL

JANET K. WILLIAMS, Representative of the American Academy of Nursing; Professor of Nursing, Chair of the Behavioral and Social Science Research Institutional Review Board, University of Iowa, Iowa City

Fellow

SAMUEL G. JOHNSON, American Association of Colleges of Pharmacy, American College of Clinical Pharmacy Anniversary Fellow

IOM Staff

ADAM C. BERGER, Project Director
SARAH H. BEACHY, Associate Program Officer
MEREDITH HACKMANN, Senior Program Assistant
ANDREW M. POPE, Director, Board on Health Sciences Policy

Reviewers

This workshop summary has been reviewed in draft form by individuals chosen for their diverse perspectives and technical expertise, in accordance with procedures approved by the National Research Council's Report Review Committee. The purpose of this independent review is to provide candid and critical comments that will assist the institution in making its published workshop summary as sound as possible and to ensure that the workshop summary meets institutional standards for objectivity, evidence, and responsiveness to the study charge. The review comments and draft manuscript remain confidential to protect the integrity of the process. We wish to thank the following individuals for their review of this workshop summary:

Alexander Djuricich, Indiana University School of Medicine
Jean Jenkins, National Human Genome Research Institute
Lucinda Maine, American Association of Colleges of Pharmacy
Michael Rackover, Philadelphia University

Although the reviewers listed above have provided many constructive comments and suggestions, they did not see the final draft of the workshop summary before its release. The review of this workshop summary was overseen by **Melvin Worth.** Appointed by the Institute of Medicine, he was responsible for making certain that an independent examination of this workshop summary was carried out in accordance with institutional procedures and that all review comments were carefully considered. Responsibility for the final content of this workshop summary rests entirely with the rapporteurs and the institution.

Acknowledgments

The support of the sponsors of the Institute of Medicine Roundtable on Translating Genomic-Based Research for Health was crucial to the planning and conduct of the workshop Assessing Genomic Sequencing Information for Health Care Decision Making and for the development of the workshop summary report. Federal sponsors are the Department of Veterans Affairs; Health Resources and Services Administration; National Cancer Institute; National Human Genome Research Institute; National Institute of Mental Health; National Institute of Nursing Research; and National Institute on Aging. Nonfederal sponsorship was provided by the American Academy of Nursing; American College of Medical Genetics and Genomics; American Heart Association; American Medical Association; American Society of Human Genetics; Association for Molecular Pathology; Blue Cross and Blue Shield Association; College of American Pathologists; Eli Lilly and Company; Genetic Alliance; International Society for Cardiovascular Translational Research; Kaiser Permanente Program Offices Community Benefit II at the East Bay Community Foundation; Life Technologies; Merck & Co., Inc.; National Society of Genetic Counselors; Northrop Grumman Health IT; Pfizer Inc.; and PhRMA.

The Roundtable wishes to express its gratitude to the expert speakers whose presentations examined the context for the challenges involved in educating health care providers in genetics, reviewed promising approaches for providing genetics education, and identified opportunities and next steps for improving genetics education for health professionals. The Roundtable also wishes to thank the members of the planning committee for their work in developing an excellent workshop agenda. The project director would like to thank project staff who worked diligently to develop both the workshop and the resulting summary.

Contents

Boxes

Abbreviations and Acronyms

AAFP	American Academy of Family Physicians
AAMC	Association of American Medical Colleges
ABIM	American Board of Internal Medicine
ACCME	Accreditation Council for Continuing Medical Education
ACGME	Accreditation Council for Graduate Medical Education
ASHP	American Society of Health Systems Pharmacists
EPA	entrustable professional activity
FDA	U.S. Food and Drug Administration
G2C2	Genetics/Genomics Competency Center for Education
IOM	Institute of Medicine
IPE	interprofessional education
ISCC	Inter-Society Coordinating Committee for Practitioner Education in Genomics
METRIC	Measuring, Evaluating, and Translating Research Into Care
MOC	maintenance of certification
RCT	randomized clinical trial

1

Introduction and Themes of the Workshop[1]

Despite the growing use of genomic applications in clinical practice, health professional knowledge about genomic information and confidence in using it have not kept pace (McInerney et al., 2012). Genetic influences on health are ubiquitous and multifaceted, which can make it difficult to use this information in a typical health care visit. Many health care providers do not have either the knowledge or the tools they need in order to apply genetic information in their day-to-day practices. This lack of support is contributing to a substantial delay in the translation of genetic research findings, when appropriate, into improvement in patient outcomes within the health care system.[2]

In addition, although the need to improve genetics knowledge among health care providers is clear, the best approaches to educating health care providers in a way that produces meaningful changes in clinical practice are not, especially given the competing coursework and training needs that exist in today's increasingly complex health care settings. Simply providing information is often not sufficient to spark interest among graduate health professional students, residents, and fellows or to

---

[1]The planning committee's role was limited to planning the workshop. The workshop summary has been prepared by the rapporteurs as a factual account of what occurred at the workshop. Statements, recommendations, and opinions expressed are those of individual presenters and participants and are not necessarily endorsed or verified by the Institute of Medicine. They should not be construed as reflecting any group consensus.

[2]In this report, the terms "genetics" and "genomics" are treated as two overlapping domains of knowledge that raise similar issues and have similar objectives. Rather than referring repeatedly to "genetics and genomics education," this report uses each term to connote the body of knowledge associated with both.

elicit behavioral change in providers (Davis et al., 1999; Forsetlund et al., 2009). But while didactic lectures and other traditional methodologies generally do not produce the desired results in terms of clinical performance or health outcomes (Davis et al., 2011; Forsetlund et al., 2009), it is not clear which other methods will be most effective in increasing engagement and interest in genetics among health profession trainees or in achieving practice change among clinicians.

To examine pragmatic approaches to improving genetics education in both graduate and continuing education of health professionals, the Roundtable on Translating Genomic-Based Research for Health of the Institute of Medicine (IOM) hosted a workshop on August 18, 2014, titled Improving Genetics Education in Graduate and Continuing Health Professional Education.[3] The workshop examined a variety of approaches that could improve the teaching of genetics in the graduate and continuing education of health professionals; these approaches included online and interactive instruction, just-in-time approaches, the development of clinical decision-support tools, and the incorporation of genetics requirements into licensing and accreditation.

The objective of the workshop was to examine the potential and the challenges of providing genetics education, to review promising and innovtive approaches to providing education to both graduate health professional students and practicing health professionals, and to identify potential next steps for achieving effective genetics education (see Box 1-1). The workshop did not address the need for evidence, which has been discussed in previous workshops by the Roundtable. Nor did it

BOX 1-1
Workshop Objectives

- To examine the context for the challenges involved in educating health care providers in genetics.
- To review promising approaches for providing genetics education in various settings.
- To identify opportunities and next steps for improving genetics education for health professionals.

[3]The workshop agenda, speaker biographical sketches, full statement of task, and list of registered attendees can be found in Appendixes A through D, respectively. For more information about the workshop, see http://www.iom.edu/Activities/Research/GenomicBased Research/2014-AUG-18.aspx (accessed September 23, 2014).

examine undergraduates or medical, nursing, or pharmacy students, but rather it was focused on those who have already graduated and are in their internship, residency, advanced degree program, or fellowship or are already practicing. The workshop considered not just physicians but also nurse practitioners, physician assistants, pharmacists, advanced practiced nurses, and others who interact with patients and need genetics knowledge. Finally, the workshop did not focus on the content of the genetics education or on how well patients understand genetics issues. Rather, the central topics discussed by participants were the policies, procedures, players, and principles that play a role in the genetics knowledge of health professionals. Within that scope, a wide range of stakeholders, including health care providers, representatives of graduate and residency programs, professional society representatives, board examiners, and education specialists, presented their perspectives and participated in discussions during the workshop.

WHAT IS THE CASE FOR GENETICS EDUCATION?

Each interested party has a different perspective on why medical education is important, said Bruce Blumberg, the institutional director of graduate medical education in Northern California for Kaiser Permanente and co-chair of the workshop. "A clinician would have one answer. A laboratorian would have a different answer. A pharmacist or a nurse practitioner yet a different answer. A student perhaps a different answer." Yet most would agree that some kind of education would improve care and yield better outcomes for patients. As an example of the difference that genetics education could make in the care that a patient receives, Blumberg presented a case study (see Box 1-2) and then asked the audience to consider the question: Who do you wish would know more about what so that patient care or outcomes would be improved?

BOX 1-2
A Case Study Presented by Workshop Co-Chair Bruce Blumberg

A 52-year-old woman presents to her family medicine nurse practitioner with a new onset of hypertension. The nurse practitioner takes a detailed family history and finds that the patient had an uncle with a kidney tumor.

The patient sees a variety of specialists, one of whom orders a DNA test. Two mutations in succinate dehydrogenase subunit B are identified. One is a mutation that has been previously identified as being causative for paraganglioma. The other is a variant of unknown significance.

A genetic counselor then orders a targeted DNA test on the 27-year-old daughter of the 52-year-old patient and finds that the mother's pathogenic mutation is absent, which reassures the daughter.

One year later the daughter presents with a neck mass that is ultimately identified as a malignant paraganglioma. A follow-up DNA analysis finds that the variant of unknown significance in the mother is also present in the daughter. The geneticist reviews the literature and questions the original assignment of pathogenicity to the other mutation.

Before starting chemotherapy, the daughter asks the geneticist to interpret her self-ordered methylation analysis to determine its implications for her therapy. The methylation analysis has nothing that resembles an interpretation.

In each step of this case study, Blumberg said, improved genetics education could have led to a different course of action, whether for the nurse, the internist, the specialists, the genetic counselor, the geneticist, or the patient.

WHAT CHANGES ARE NEEDED, AND HOW CAN THEY BE MADE?

What are the three things that would have the greatest impact in improving genetics education for health care professionals? Joan A. Scott, chief of the Genetic Services Branch in the Maternal and Child Health Bureau of the Health Resources and Services Administration and workshop co-chair, asked all meeting participants to consider this question from their own particular perspectives as the workshop progressed. What are the provider education goals that need to be achieved, she asked, and how can this be done? Who needs to be targeted, and what specifically do we need them to do?

Essentially, Scott said, it is necessary to identify the change that needs to be made and the way to make that change occur. The answer may differ from one part of the health care system to another, she said, but it is possible to identify several broad themes. First, there is widespread agreement within the health care community that genetics and genomics are indeed relevant to clinical practice. This agreement can be a powerful force for change once the needed changes have been identified along with the steps that need to be taken to realize those changes. (Chapter 6 summarizes the suggestions made by workshop participants about needed changes and next steps.)

Another broad theme that Scott identified was the importance of interprofessional education (IPE) as a way of increasing genetic literacy across the health care community. Different kinds of health care providers will need to interact with each other in educational and practice settings to integrate genetics knowledge into mainstream clinical practice.

Finally, Scott said, these interactions need to extend beyond health care providers to the many other groups that are involved in medical education. For genetics knowledge to inform medical practice in a way that improves public health, groups that accredit educational programs, that design and evaluate new educational approaches, and that focus on quality improvement within health care must all work together. "It isn't until we are all together talking about this as a holistic issue that we are really going to be" making that particular change, she said.

ORGANIZATION OF THE REPORT

Chapter 2 provides an introduction to the major topics of the workshop by looking at some of the myths and mistakes surrounding graduate medical education and continuing medical education and at how those misconceptions can lead to problems in the field. Chapter 3 examines several promising new approaches to medical education, along with several of the broad issues associated with that education.

Chapter 4 examines graduate-level genetics education for health professionals, including in internships, residencies, and fellowships for physicians and the equivalents of those stages for other health care providers. Chapter 5 looks at continuing professional education and at the competencies that this education is designed to develop.

Finally, Chapter 6 asks how to make the necessary changes in graduate and continuing health professional education using the policies, procedures, players, and principles discussed earlier in the workshop.

2

Myths and Mistakes in Graduate and Continuing Medical Education

Teaching about genetics and genomics requires touching on a broad range of issues that are relevant throughout the health care system because these disciplines bring together the science of medicine, patient centeredness, ethical decision making, and commercial support, said David A. Davis, the senior director of continuing education and performance improvement with the Association of American Medical Colleges (AAMC). Davis described some of the myths and mistakes in graduate and continuing professional education in medicine and traced how these myths and mistakes contribute to a serious problem in the health care system: the gap between what is known by medical researchers about genetic influences on health and how this knowledge is applied in current medical practice.

EDUCATION MYTHS

A number of myths have been perpetuated about educating health professionals, Davis said. The first myth is that it is always good to provide people with more information. When there is no information, giving information to individuals tends to improve their performance. However, once an optimal point is reached, improvements tend to decline or disappear, Davis said, citing the Yerkes–Dodson Law (Yerkes and Dodson, 1908).

The second myth is that education only occurs through lectures or by attending conferences. But health care providers "learn every hour of every day," Davis said, not just in conferences or courses. A related myth is that lectures change behavior. Didactic sessions largely do not lead to improved professional practice (Forsetlund et al., 2009). Behavior changes are much more likely to be elicited by audit and feedback or by more interactive sessions, Davis said.

A fourth myth is that continuing medical education is only for physicians or about the receipt of credit. In fact, all health care providers need such education, Davis said, and they actually have many reasons for participating in educational activities.

Finally, it is a myth that continuing medical education and, to some extent, graduate medical education, are isolated activities unrelated to health systems, health care delivery, or patient outcomes, Davis said. "If we think about it that way, we are bound to fail in genomics education."

THE CLINICAL CARE GAP

Today, a gap exists between the knowledge available to inform health care and what actually happens in health care settings, Davis said. That gap may manifest itself in a lack of information about a fast-spreading infectious disease, the errors that occur in emergency rooms, the proper use of diagnostic imagers, doctor–patient communications, health disparities among counties or states, or the difference between ideal evidence-based practice, as represented by the many published guidelines available to doctors, and current practice, as represented by the paper records in many doctors' offices.

The clinical care gap, Davis said, has several origins, including the evidence supporting or the guideline directing a practice, the educational delivery system, clinician training, and issues with the health care system. Clinicians can be overwhelmed by the amount of information that exists. The national guideline clearinghouse maintained by the Agency for Healthcare Research and Quality has approximately 3,000 sets of guidelines.[1] For each, clinicians must consider such factors as the relative advantage of a guideline to a clinician and a patient, the compatibility of a guideline with practice, its complexity, its cost, its observability in practice, and its sustainability. Guidelines must also be carefully thought about, Davis said, because they do not traditionally incorporate experiential evidence.

[1]See http://www.guideline.gov (accessed November 11, 2014).

Clinicians undergo many years of training, Davis said, and some forms of education are more effective than others. For example, flipped classrooms, in which students listen to lectures outside the classroom and work on specific problems with their instructors, have been shown to be more effective in many cases than traditional lecture-based classes. Effective education may take the form of predisposing individuals to change, as well as enabling and reinforcing that change. As is the case with people in general, physicians are not very self-aware, Davis said. "We do not know what we do not know. Thinking about education, absent things like feedback and observable data, means that learning may fall fallow."

Finally, many complex aspects of the health care system influence the translation of evidence to practice, including the roles of health care teams, the availability and utilization of resources, governance, and leadership.

CONTINUING MEDICAL EDUCATION AND THE CLINICAL CARE GAP

Does continuing medical education, Davis asked, change the behavior of physicians and improve health outcomes? It can, he said, but many of the myths surrounding and mistakes present within continuing medical education can contribute to the clinical care gap.

A major portion of continuing medical education fails to reflect the research literature on educational effectiveness, Davis said. For example, the research shows that providing knowledge is necessary but not sufficient for change. Rather, as discussed earlier, effective education often involves predisposing people to change, enabling change, and reinforcing change in addition to providing knowledge. Other results of education research also bear on medical education, Davis said. For example, approaches that elicit interactivity are better instructional tools than lectures; these approaches include case discussions, reflection, and audience-response systems. Needs assessments followed by performance feedback are also important to effective instruction. And learning can occur in many places—online, from patients, from reminders at the point of care, from simulations, from small-group learning activities, from algorithms and care pathways, and from today's ubiquitous "smartphones and tablets."

Finally, continuing medical education and graduate medical education do not occur in isolation. Many factors influence them, including accreditation requirements, quality improvement initiatives, new technologies, the information explosion, the Affordable Care Act, and benchmark techniques.

Continuing medical education is a very large enterprise, Davis noted. According to data from the Accreditation Council for Continuing Medical Education, physicians participated almost 14 million times and allied health professionals participated more than 9 million times in 2011,[2] participating in 953,000 hours of instruction and 133,000 activities. Yet, despite all this activity, the clinical care gap persists, Davis said. "We need to think about better ways to do it."

LESSONS LEARNED

The framing and language of the message that is wanted to be conveyed need to be carefully thought out, Davis said. Is a 70-page guideline needed, or could it be a simple guidance that is consistent with current practices? The relative advantage a guideline might have, its compatibility with practice and complexity, the added costs to providers and patients, and the sustainability of the practice over time should be considered, Davis said.

More effective means of education are available, Davis said, and they need to be used (IOM, 2010). Just-in-time learning, for example, enhances interactivity and provides information that is likely to be more relevant to the task at hand. (Discussed in detail in Chapter 3.)

All health professionals should be involved in continuing medical education, not just physicians, Davis said. In addition, the staging of educational innovations or interventions needs to be considered. For example, evidence-based practice can be conveyed through role models rather than didactically, although ensuring a sufficient supply of faculty may be a problem.

Leverage change, Davis suggested. Identify other things that are occurring in the health care system that may be able to reduce the clinical care gap and leverage those to elicit the desired change. For example, Davis mentioned an implementation planning guide for continuing medical education called the PROCEED model that is based on the idea of

---

[2]Individuals might attend multiple activities during a year, thus these numbers do not reflect the total number of health professionals that participate in continuing medical education; http://www.accme.org/news-publications/publications/annual-report-data/accme-annual-report-data-2011 (accessed November 11, 2014).

predisposing an individual to change, enabling that change to occur, and reinforcing that change (Davis et al., 2003). Predisposing, enabling, and adopting might involve workshops, small-group learning sessions, or just-in-time learning activities. Adherence might involve reinforcement through such means as audits, feedback, and point-of-care reminders. Another existing resource is the MedEdPORTAL website (www.mededportal.org) of AAMC, which provides genetics teaching resources. Finally, as in all areas of education, messages can be spread by embedding them within current education and practice approaches and by using a train-the-trainer model. "We do not need to re-create the wheel," Davis concluded.

3

Educational Approaches

Important Points Highlighted by the Individual Speakers

- Innovations such as just-in-time and online education can get health care professionals the information they need when they are most likely to be able to use and learn it.
- Technological advances can be a powerful adjunct to, but not a replacement for, effective instruction and the acquisition of clinical skills.
- Interprofessional education can achieve better communication and collaboration among learners, practitioners, patients, clients, families, and communities.
- Avoiding conflicts of interest in educational materials requires a strong commitment by the leadership of the medical community.
- Implementation research can promote the integration of research findings into routine use in an evidence-based manner.
- The motivations of practitioners and the attributes of educational resources are key determinants of whether genetics education will be embraced.

New research findings and evidence-based approaches to education have transformed many learning environments and have improved learning outcomes. Workshop speakers discussed several of these new approaches, including just-in-time approaches to education, the use of social media and other technology platforms in health professional education, IPE, and the use of evidence and theory in designing educational activities, along with a number of ongoing issues in medicine and genetics education.

JUST-IN-TIME APPROACHES TO EDUCATION

Disease-specific genetic knowledge is changing rapidly, with new variants being identified every day, said Benjamin Raby, associate professor of medicine at Harvard Medical School and the genetics section editor for UpToDate. "We need to get that information and figure out how to integrate it in a rapid way," he said, so that patients are receiving the best care possible. However, in his role as the director of a pulmonology genetics center who often sees patients with extremely rare disorders, Raby has witnessed firsthand what he called the "dilemma" of providing genetics education. Late-stage fellows and junior faculty in his training program at Harvard Medical School, who should already have a solid training in genetics, are not able to answer such basic questions as what is a haplotype, a genotype, or an allele, or what genetic tests should be ordered in a given situation, or whether screening should be recommended for relatives. "We have to think about the fundamental basic genetic principles in our education," Raby said, and also be concerned with how to convey variant-specific and disease-specific information.

Just-in-time approaches to education offer a potential way to deliver an array of information at exactly the point in time at which it is most needed or useful. UpToDate is an online, subscription-based, just-in-time clinical decision support resource, which continuously reviews the medical literature in order to provide clinicians with the information they need to make medical decisions. The UpToDate contributors review approximately 460 medical journals covering more than 10,000 topics in 21 specialties. Additions to the UpToDate resource undergo a three-tiered peer-review process—first by independent reviewers, then by a section editor, and finally by editorial staff. The resource is used by more than 850,000 clinicians from 29,000 institutions and practices in 164 countries, with 90 percent of U.S. academic medical centers having subscriptions. In 2013, it received more than a quarter billion queries, with an average of 484 topic reviews per minute. And its use has been associated with "demonstrable improvements in quality of patient care," Raby said (Isaac et al., 2012).

New knowledge is continually being incorporated into UpToDate, he added. The section editors examine the tables of contents from every journal on a monthly basis to identify articles that might be worthy of further review. They also respond to user suggestions for topics that could be included or revised, Raby said. Authors typically update topics once per year, but pressing topics are revised more frequently.

In the area of genetics, UpToDate seeks to provide two types of content—disease-specific and foundational. Disease-specific content includes information about specific monogenic and polygenic disorders as well as polygenic contributions to common diseases plus other disease-specific topics. Current content includes such specific topics such as breast and ovarian cancer, early onset Alzheimer's disease, cardiomyopathy, and colorectal cancer, among many others. The types of variation, the genes considered to be relevant, what to tell a patient, and the next line of testing are all discussed within the disease context. Foundational content, by contrast, includes basic information such as a glossary of genetic terms, basic principles of genetic disease, an overview of genetic variation, and information on incidental findings. Entries also include hyperlinked references—for example, from specific disease context to related foundational information.

The genetics section is unique within UpToDate, Raby said, in that other sections do not have similarly broad treatments of foundational issues. But the need for educational material in genetics is so pressing that the genetics section tries to provide that framework. At the same time, genetics information is included with content related to other specialties because this is seen as an effective way of increasing understanding of genetics.

The UpToDate resource tries to make information as accessible as possible by having the information directly linkable from electronic medical records, Raby said. UpToDate is also currently working on integrating calculators of risk prediction and decision-management solutions into electronic medical records, which could be pre-populated with patient-specific information.

Raby concluded by pointing to several remaining deficiencies in the UpToDate resource. One is the lag time in providing updated information because of the rapid evolution of genetic information. "New knowledge is always being developed around variants that have already been reported out," Raby said. "What do we do when you get back a report that says, 'This is likely benign,' and 3 years later it turns out we made a mistake?" Furthermore, simply listing an identified gene variant on a report is of little value to clinicians. There is a need for supported reports that include the evidence for an interpretation and that tell clinicians specifically how to think about a given result. Such reports provide not simply data but knowledge, Raby said.

Finally, the updating and sharing of genetic information remains a challenge. New systems are being developed to link genotypic and phenotypic information across laboratories without violating privacy concerns. "If I see a new variant, and it turns out that it has never been published on before, but four other labs have seen the same rare variant and the same phenotype is linked, we can now make much better inferences," he said. Such information can provide the basis for a continually updated clinical portal, with improved communications between laboratories and physicians and continuous education. "These are the types of tools that we need to get into place for people to have the knowledge they need to move forward."

TECHNOLOGY IN MEDICAL EDUCATION

Emerging technologies and media platforms offer other new and intriguing ways to educate students and health care providers in genetics. But technology needs to serve patients as well as learners, said Alexander Djuricich, associate professor of clinical pediatrics and clinical medicine at the Indiana University School of Medicine. "I am interested in emerging technology for myself, my learners, and patients for the purpose of improving education and ultimately patient care," he said.

The potential of technology as a learning tool has been widely recognized. Technology can improve engagement with learners, provide educational opportunities outside of traditional lecture room settings, and can foster the wider dissemination of information, Djuricich said. These attributes have led a number of institutions and organizations to adopt technology for use in medical education. Djuricich said that the University of California, Irvine, has been giving all of its incoming medical students iPads, for example. An asynchronous global emerging medicine journal club is now being hosted on a blog by the *Annals of Emergency Medicine*, with comments moderated on the blog and Twitter (Radecki et al., 2014), and audience response systems are becoming more common, the use of which has been shown by research to improve classroom engagement and learning (Forest, 2012).

Twitter can be especially effective in conferences, Djuricich said, because it can act as a back channel for communication with the people at the meeting and elsewhere. Tweets can highlight important aspects of a meeting and can even be done as part of a formal presentation (Chaudhry et al., 2012; Desai et al., 2012; Logghe et al., 2013; Matta et al., 2014; Neill et al., 2014). For example, Djuricich's institution has piloted tweeting grand rounds for pediatrics. In 17 presentations, 613 separate tweets were sent from 61 participants. "This provides a great example of the power of disseminating information, not just in the room for that 1 hour, but to disseminate it to others who can either read it then or at a later time."

Technology can be very effective in making information useful and meaningful for clinicians and patients, but its use must be balanced with still demonstrating humanity, Djuricich said. At the same time as the journal *Pediatrics* opted to move more toward a digital identity by developing a full-text mobile app and becoming active in social media tools, such as Facebook and Twitter, in response to reader preferences for accessing journal content (First et al., 2014), an article was published in *Academic Pediatrics* warning against the overuse of technology and the effects of such overuse on interactions with patients and learners (Crain, 2014). Similarly, Colbert and Chokshi (2014) and Toll (2012) have cautioned about the potential drawbacks of technology. As the former wrote, "Technology itself is no panacea. . . . Some fear that [learners] may be losing sight of humanism in medicine" (Colbert and Chokshi, 2014). We need to be careful about how we use technology, and we need to provide instruction on how to use it effectively, Djuricich said.

Emerging technologies are often treated as a tool by which to teach content. But technology itself can also be the content, as is the case in a workshop on how to use social media or a session on the best way to use audience response systems such as clickers. As an example of this approach, Kind et al. (2014) have published a set of tips on using social media as a medical educator. They include

- Identify and reflect on one's digital identity.
- Observe and establish comfort first—"Think, then contribute."
- Use social media to disseminate evidence-based health information.
- Apply social media guidelines, such as those developed by academic medical centers and professional groups.
- Tap into the power of community, whether students, physicians, medical educators, or patients.

Gaglani and Topol (2014) have urged that training programs and medical schools integrate mobile health technologies into their curricula. Djuricich agreed that this is a promising approach and argued that this education should extend to continuing professional development as well, but he also emphasized that while technology can be effective as an adjunct to instruction, it should not be seen as a replacement for effective instruction. Not all technologies are ready for use, and it can be difficult for clinicians to keep up with their development. Technologies tend to go through a hype cycle of inflated expectations, disillusionment, gradual enlightenment, and, finally, productive use.[1] For example, new apps are being created so quickly that no one can keep up with all of them. But apps need to be evaluated, which has led to the development of rubrics and websites, such as imedicalapps.com and happtique.com, to assess the value of apps. Such apps are also subject to U.S. Food and Drug Administration (FDA) regulation if they can affect a user's health, though so many products are currently being developed that FDA is hard pressed to keep up.

Technology can augment but not replace the acquisition of clinical skills, Djuricich concluded. In particular, it has the potential to increase engagement and to reach learners wherever they may be.

INTERPROFESSIONAL EDUCATION IN GENETICS

IPE is an approach to education in which individuals from two or more professions learn about, with, and from each other to improve collaboration and the quality of health care. "About, with, and from" are the three key concepts, said Diane Seibert, professor and chair and director of the Family Nurse Practitioner Program at the Uniformed Services University of the Health Sciences. The concept of IPE is to learn *about* what another profession does, *with* other professionals in the same environment, and taught *from* different professions. This can be done, for instance, in a faculty-led or small-group environment where medical and nursing students are trained together, Seibert said.

The goal of IPE is to improve collaborations between health care professionals in order to improve patient outcomes. Interprofessional learning may be spontaneous or strategically embedded in an educational program, but the ultimate goal is better communication and collaboration

[1]See http://www.gartner.com/newsroom/id/2819918 (accessed January 16, 2015).

among learners, practitioners, patients, clients, families, and communities. "If people communicate better with one another, they will practice as a team, and you will get better patient outcomes," Seibert said. "You have a diverse group of people bringing their perspectives to a common place and sharing and collaborating."

IPE is useful because health care systems are so complex that different perspectives are needed to get the best outcomes, Seibert explained. In addition, IPE fosters mutual respect among disciplines, countering the negative stereotypes about other disciplines that can begin early in professional education and shape how people view other professions for the rest of their careers.

Health educators have been interested in IPE for a long time. The 1972 Institute of Medicine report *Educating for the Health Team* concluded that "a major deterrent to our efforts to fashion health care that is efficient, effective, comprehensive, and personalized is our lack of a design for the synergistic interrelationship of all who can contribute to the patient's well-being" (IOM, 1972, p. 4).

More than 40 years later, IPE is still not where it needs to be, Seibert said. There are still not a lot of opportunities for students to learn in this manner. The Interprofessional Education Collaborative Expert Panel (2011) recently published *Core Competencies for Interprofessional Collaborative Practice*, which has been "getting some forward traction," according to Seibert. For example, in October 2012 the Liaison Committee on Medical Education approved a new accreditation standard,[2] which went into effect in July 2013, that requires that "the core curriculum of a medical education program must prepare medical students to function collaboratively on health care teams that include health professionals from other disciplines as they provide coordinated services to patients. These curricular experiences include practitioners and/or students from other health professions." As a result of this new standard, students will have more opportunities to interact with other disciplines before they begin their professional careers, which in turn could have an effect on continuing medical education.

While IPE is beginning to be more broadly embraced, it continues to face challenges, Seibert said. Not much evidence yet exists to show that IPE is more effective than standard approaches in terms of learning out-

---

[2] See http://www.lcme.org/connections/connections_2014-2015/ED-19-A_2014-2015.htm (accessed November 11, 2014).

comes or health outcomes. Longitudinal research will be needed to demonstrate such effectiveness, Seibert said. In addition, developing a curriculum for very diverse audiences can be challenging. Faculty who have not had these experiences may not know how to develop such curricula or how to facilitate small interprofessional groups. "Quite a bit of faculty development is needed in this area," Seibert said.

Scheduling poses other difficulties. For example, different professions tend to hold their conferences at different times of the year, making it difficult for members of various disciplines to meet together, Seibert said. "We need to think about how do we break those silos down and get multiple disciplines together in the same room. [Currently] there is no cross talk."

Genetics and genomics are cross-cutting topics that could help break down those barriers and attract members of various health care professions, Seibert said. Already, patients with rare single-gene disorders tend to be treated through team-based approaches. However, many members of the health care community have not had much experience with genetics and genomics because of limited educational experiences and the rapidly advancing state of the field. As a result, while the audience for genetics education is diverse, its members tend to have similar learning needs, which could address some of the challenges of developing an interprofessional curriculum.

People will be most interested in IPE events that meet their needs, Seibert said. "The topic will attract the audience." For example, nurse practitioners are very interested in family health histories and risk assessment. Many individuals are also interested in genetic testing, especially testing for common complex diseases that affect large numbers of patients. A rare disorder that health care professionals are unlikely to see in their practice may not be particularly attractive. Members of many different professions are also interested in ethical, legal, and social issues or counseling because many professionals have relatively weak skill sets in those areas, Seibert said.

Diversity at every level is key in such events, Seibert said, and this includes having diverse audiences, diverse faculty members, diverse panels, and diverse small groups. Advertising such an event requires care, because genetics can scare people away. Not mentioning disciplines will help keep people from excluding themselves. And more complex case-based approaches can attract audiences, Seibert said, while at the same time providing an opportunity to teach participants about less common disorders or topics.

Techniques to foster engagement, including problem solving, peer instruction, advanced simulations, and built-in interactions, can enhance the success of such events. Another critical element of an interprofessional learning opportunity is allowing everyone to have a voice—for example, by including more than one representative of a profession within each group so that people are more encouraged to speak and represent multiple viewpoints within a single profession. An emphasis on building professional networks across communities can also increase engagement. In this way, participants get more out of an educational experience than just information, Raby said. "They get to meet people who they will be encountering often, maybe over their practice lifetime." Raby suggested that individuals could be further engaged by shadowing other professionals in their clinics. "Hands on is [a] really important means of learning from colleagues," he said. This could be incentivized within the hospital system, he added. Maren Scheuner, chief of medical genetics in the Veterans Affairs Greater Los Angeles Healthcare System, added that focusing on a specific condition and finding the venue where "everyone gets together" is a valuable way of facilitating IPE. Attending and regularly participating in tumor boards has been helpful in implementing Lynch syndrome screening in her region, she said.

Finally, collecting data during and after IPE events can advance the science and help achieve the goals of IPE, Seibert said. Such research can improve teamwork and facilitate the ultimate goal of improving patient outcomes.

IDENTIFYING AND AVOIDING CONFLICTS OF INTEREST IN EDUCATIONAL MATERIALS[3]

Many physicians have a misplaced confidence that marketing bias is easy to recognize and straightforward to avoid, said Jean Silver-Isenstadt, executive director of the National Physicians Alliance. That belief makes it easier for marketers to embed themselves in medical education, research, and practice at many levels. But these marketing efforts to influ-

[3]For further information on managing conflicts of interest and facilitating innovation, see: http://www.nap.edu/catalog/18723/conflict-of-interest-and-medical-innovation-ensuring-integrity-while-facilitating (accessed January 16, 2015) and http://www.nap.edu/catalog/12598/conflict-of-interest-in-medical-research-education-and-practice (accessed January 16, 2015).

ence physicians are "a destructive force," Silver-Isenstadt said. They can lead to the unnecessary use of tests, drugs, procedures, and devices, driving up costs that could be avoided. They also can direct curiosity and research away from preventive measures and other promising approaches that do not yield profits.

The National Physicians Alliance was founded in 2005 to strengthen physicians' civic engagement on behalf of patients and to build community within the profession. As part of its efforts to help restore trust and integrity in medicine, it accepts no funding from pharmaceutical or medical device companies. "The founders [of the alliance] felt that medicine was losing its way and that what had been a profession was becoming more and more grounded in business interests," Silver-Isenstadt said. "There was real risk and loss in that."

The founding campaign of the National Physicians Alliance, which continues today, is called the Unbranded Doctor. Silver-Isenstadt described it as a consciousness-raising effort that includes such elements as support for the Physician Payments Sunshine Act, which has created a publicly searchable database of gifts and payments from industry to providers,[4] opposition to the sale of individual prescriber data to drug companies for marketing purposes, and encouragement for clinicians to avoid free lunches, samples, visits from drug company representatives, and so on.

Another core effort of the alliance is the Promoting Good Stewardship in Clinical Practice project, which is funded by the American Board of Internal Medicine Foundation. It was designed to champion evidence-based care, and it sparked the Choosing Wisely campaign,[5] which promotes the ability of patients to choose care that not only is supported by evidence but is not duplicative with other tests that have been performed previously, that is free from harm, and that is necessary for their care. One aspect of the Promoting Good Stewardship project was the creation of "top five" lists of common practices within internal medicine, pediatrics, and family practice that are not supported by clinical evidence and guidelines, such as prescribing antibiotics for a viral infection, rushing to order a computed tomography (CT) scan at the first sign of uncomplicated lower-back pain, or prescribing a brand-name statin rather than a generic—"things that people know better [than to do] but happen routinely," Silver-Isenstadt said. The Choosing Wisely campaign has

[4]See https://openpaymentsdata.cms.gov (accessed January 16, 2015).
[5]See http://www.choosingwisely.org (accessed November 11, 2014).

helped expand the development of these lists—at the time of the workshop, more than 60 specialty societies had generated lists—which have reached millions of people. "It is an inspiring story because it forces all of us to consider the question of why, across specialties, people are doing so many things that are counter to the evidence," she said.

Perry Pugno, the senior advisor for education for the American Academy of Family Physicians (AAFP), observed that today's gold standard therapy may, in the future, subsequently be found to be detrimental to the health of patients. "This context should make us humble in our biases and how strongly we feel about them," he said.

Disclosure is a common way of dealing with conflicts of interest, but, Silver-Isenstadt said, while disclosure is necessary for addressing conflicts of interest, it is not sufficient. Disclosure can suggest that someone has special expertise, it can heighten trust, and it can increase bias in advice (Cain et al., 2005). Some people point to disclosure as an obstacle to meaningful reform because it is seen as the solution to the problem. As a result, time is spent trying to tweak disclosure forms, causing frustrations for those who have to fill out the forms, Silver-Isenstadt said, rather than spending time on "the real problem, which is what is behind what is being disclosed."

When making consumer decisions, people routinely consult independent evaluations of products' safety, cost-effectiveness, and price. Yet physicians routinely allow those with a personal stake in drugs, tests, and devices to fund and conduct continuing medical education, to ghostwrite journal articles, to make sales visits, to set up paid consulting agreements, and to distribute free products, Silver-Isenstadt said.

Marketing-free education requires a strong cultural commitment by the leadership of the medical community. "Vigilance about conflict of interest needs to be built into the workflow," Silver-Isenstadt said. What policies and safeguards are in place? How can they be stronger? Are they built into compliance training so that people are familiar with them? Are they well enforced? Are there consequences? Are they celebrated by the leadership? Are they points of pride?

Resources are available to address the issue. The American Medical Student Association has rated medical schools with letter grades across domains on the strength of their conflict-of-interest policies, and the association's new scorecard will rate teaching hospitals.[6] The Pew Charitable

[6]The rating is available at www.amsascorecard.org (accessed November 11, 2014).

Trust has issued best-practice recommendations for academic medical centers, including such topics as pharmaceutical samples, ghostwriting, and meals sponsored by industry.[7] Community Catalyst has put together toolkits for creating a new conflict-of-interest policy or strengthening an existing policy at a medical school or academic medical center.[8] The National Physicians Alliance has an online archive of a webcast known as the Grand Rounds series, which addresses conflict-of-interest issues.[9]

Internal conflicts of interest are pervasive as well, a participant noted. These can be manifested in fee-for-service clinical medicine or in the incentives around career advancement.

All people have personal perspectives and biases, Silver-Isenstadt said, and "to think that we could ever shed those is delusional." But people need to recognize that they are susceptible to influences and be vigilant for the sake of patients. "For us to allow the conflation of marketing and education is to abdicate a responsibility to patients," Silver-Isenstadt concluded. "And to fix it is going to require a lot of courage."

PRINCIPLES OF EVIDENCE IN DESIGNING EDUCATIONAL PROGRAMS

Educational programs fall under the rubric of implementation research, which has been defined by the Fogarty International Center as "the study of methods to promote the integration of research findings and evidence into health care policy and practice,"[10] as Maren Scheuner of the Department of Veterans Affairs explained. Similarly, Rubenstein and Pugh (2006) have defined implementation research as "scientific investigations that support movement of evidence-based, effective health care approaches (e.g., as embodied in guidelines) from the clinical knowledge base into routine use."

---

[7]The recommendations are available at http://www.pewtrusts.org/en/research-and-analysis/reports/0001/01/01/conflictsofinterest-policies-for-academic-medical-centers (accessed November 11, 2014).

[8]Information is available at http://www.communitycatalyst.org (accessed November 11, 2014).

[9]The series is available at http://www.npalliance.org/conflict-free (accessed November 11, 2014).

[10]See http://www.fic.nih.gov/ResearchTopics/Pages/ImplementationScience.aspx (accessed November 14, 2014).

The success of implementation research can be measured using two methods, Scheuner said. In formative evaluations, a rigorous assessment process is designed to identify potential and actual influences on the progress and effectiveness of implementation efforts (Stetler et al., 2006). Such evaluations can use both qualitative and quantitative methods, including semi-structured interviews, direct observation, surveys, and tools that assess the fidelity of adoption of the implementation strategy. Formative evaluations can also provide much better understanding of the variation seen in the uptake of guidelines. In summative evaluations, by contrast, a systematic process is undertaken of collecting and analyzing data on impacts, outputs, products, outcomes, and costs in an implementation study.

In conducting implementation research, the use of theory can help in planning the implementation strategy, aid in the conduct of evaluations, identify unanticipated elements critical to successful implementation, and help in understanding findings, including relationships between domains or constructs. Additional insights can also be gained about the theory itself, which can benefit future studies. There are three basic types of theories: explanatory theories, which include hypotheses and assumptions about how implementation activities will facilitate a desired change as well as about the facilitators and barriers for success; process theories, which describe how implementation should be planned, organized, and scheduled; and mixed theories, which include elements of both explanatory and process theories. Often, multiple theories will be used in implementation research, Scheuner said.

Scheuner has used these methodologies—formative and summative evaluation as well as theory—to examine genetic testing applications that have evidence of clinical utility, particularly with regard to family history. The project, known as Family History Education to Improve Risk Assessment for Hereditary Cancer, sought to improve the recognition and referral of patients at risk for hereditary cancer syndromes by primary care clinicians at women's health clinics. The program reflected research showing that sequenced, continuous, and multifaceted activities can lead to a change in practice (Mazmanian and Davis, 2002) and that multifaceted interventions targeting different barriers to change are more likely to be effective than single interventions (Grimshaw et al., 2001). It also incorporated research indicating that provider behavior is influenced by a number of factors, including provider characteristics, organizational structures and processes, external environmental factors, patient characteristics, and the encounter between the patient and the clinician (Rubenstein et al., 2000).

During the pre-implementation phase, the formative evaluation included recognizing determinants of current practice, identifying barriers and facilitators, assessing the feasibility of the proposed intervention, and integrating findings into the intervention design and refining the design prior to implementation. This evaluation revealed that the electronic health records lacked standards for family history documentation. Interviews with primary care clinicians at the facility revealed that the providers wanted a standardized template for the family history in the electronic health record, better organization of the family history in the record, and data gathered from patients before their office visits through a kiosk, personal health records, or some other method. The providers gave a high rating to clinical reminders that could help with documentation, would enable easier recognition of conditions which might be inherited, and that prompt referrals for consultation or testing. When asked the reason for giving high ratings to such reminders, the providers pointed to, among other things, a lack of knowledge, familiarity, and confidence in genetic risk assessment, diagnosis, and testing. "Most primary care clinicians do not want any reminders," Scheuner said, "but they were open to it [in this case] because they knew they needed help to better document family history."

In response, Scheuner and her colleagues created a draft reminder tool designed to identify family history red flags for hereditary cancer, eventually expanding the reminder template to gather even more comprehensive family history information. Ultimately, a multifaceted strategy that included clinical interventions, information interventions, and behavior interventions was developed as part of a family history toolkit. For example, the strategy included a lecture series, information sheets for providers and patients, a website with additional information, and quarterly practice feedback reports.

With the formative evaluation, project developers sought to assess discrepancies between the implementation plan and its execution, to understand and document the nature and implications of local adaptation, to monitor impacts and indicators of progress toward project goals, to use data to inform the need for modifying the original strategy, and to provide positive reinforcement to high performers and negative reinforcement to low performers.

The project led to a doubling of cancer family history being documented in progress notes, Scheuner said. Missed opportunities to make genetics referrals also decreased over time. Individualized reinforcement, both positive and negative, was given to providers.

Post-implementation interviews with clinicians sought to assess the usefulness and value of the intervention from the stakeholders' perspectives, to elicit stakeholder recommendations for further intervention refinements, to assess satisfaction with the intervention and implementation process, and to identify additional barriers and facilitators. Almost all of the clinicians interviewed said that they were doing a much better job of documenting the cancer family history and that they were much better informed about when to make a referral, Scheuner said.

Effective programs tend to be multifaceted and ongoing, Scheuner concluded, and the use of theoretical models can help inform the development and evaluation phases for such programs. Djuricich added that much could also be learned from examining the partnerships that clinicians in specialties that do not have enough practitioners to meet health care needs, such as adolescent medicine, child psychiatry, and geriatrics, enter into with primary care organizations in order to disseminate education. Investigating how well these partnerships work might be a way to understand how to advance genetics education. Scheuner said that there is a need to better understand what works in different health settings and systems in order to identify what works locally and what could work nationally. "I would encourage you to think about developing implementation strategies for your genetics education programs and generate evidence about the effectiveness," she said.

MEASURING THE EFFECTIVENESS OF EDUCATIONAL INTERVENTIONS

Much can be learned throughout the course of an intervention, rather than waiting until the intervention has been completed, Scheuner said. In projects at the Department of Veterans Affairs, ongoing interviews are conducted with stakeholders, including front office staff. In particular, she emphasized the importance of quality improvement and the use of best practices, which requires quality indicators. "Everybody needs to agree on that," she said. "Then you have to go and look. If you can capture them in administrative data, that might be great. Oftentimes it is not that simple."

Djuricich pointed out that educational researchers have studied the impact of educational interventions—for example, by using the seven levels developed by Moore et al. (2009):

1. Participation
2. Satisfaction
3. Learning
4. Competence
5. Performance
6. Patient health
7. Community health

Murray Kopelow, the president and chief executive officer of the Accreditation Council for Continuing Medical Education, noted that numerous systematic reviews have been performed that demonstrate that continuing education does exactly what it is intended to do (Al-Azri and Ratnapalan, 2014; Bloom, 2005; Davis and Galbraith, 2009; Forsetlund et al., 2009; Lowe et al., 2009; Mansouri and Lockyer, 2007; Marinopoulos et al., 2007; Mazmanian et al., 2009). The issue, he said, is "that professionals do not do what they are supposed to do. This is a barrier." Putting resources into the further evaluation of continuing medical education initiatives would not be well spent. Instead, he suggested doing research on the problems that exist, such as the professional practice gap, and how those problems can best be addressed.

Davis emphasized the importance of having a gold standard to provide a measure against which performance can be gauged. "We need a guideline that says, in this disease, for this condition, with these parameters of age or gender, we need to do this test. . . . Then we can do measurements in population health standards." He also said that national campaigns like Choosing Wisely, combined with local agreements to adopt a particular guideline, can be means to facilitate changes in practice.

REACHING COMMUNITY PRACTITIONERS

A number of factors can influence the motivation that community practitioners have for embracing and using a new practice approach, Perry Pugno said, and these factors include

- It directly benefits patients the practitioners see frequently.
- It is simple and straightforward to apply.
- It reduces practice costs or saves time.
- It reduces professional liability risks or costs.
- It helps secure privileging.

- It is part of a maintenance-of-certification (MOC) requirement.
- It leads to financial remuneration.

Pugno emphasized a point made in the previous chapter: Giving clinicians information does not necessarily mean that they will integrate a new approach into their practice "People do not learn by just telling them things," he said. "We have to make them do something. They have to actually use the information for them to integrate it and make it part of what they do on a daily basis."

Adoption and use are further complicated when dealing with genetics information, Pugno said, because of the gap between the promise of genetics and the current reality. Genomics was touted as making it possible to deliver targeted care with treatments that would have a higher rate of success and also to make it possible to use genetic information to predict the risk of or prevent the occurrence of disease. However, the majority of applications for genetics have been focused on rare disorders, they are infrequently relevant to practice, and can be an expense to the patient or practice with little or no benefit currently.

AAFP has taken several steps to equip its 100,000 practicing physicians with information and skills that they will embrace and use. For example, it has developed a MOC module known as Measuring, Evaluating, and Translating Research Into Care (METRIC), which enables the capture of performance information in order to improve patient care and outcomes. A recent study on the efficacy of METRIC found that in nearly 12,000 patient encounters with family physicians who had undergone the intervention, 75 percent of the quality markers in diabetes care showed improvement (Bird et al., 2013). AAFP also has developed such tools as self-evaluation modules with online patient simulators, fast-track board review with audience response and engagement systems, a translation-to-practice credit system, geospatial mapping of needs assessments, and the use of social media to facilitate interaction. "There are a lot of efforts that are going on right now to try to provide education that matters that clinicians will accept and use in the care of their patients," Pugno said.

If educational resources are to help capture the interest of practicing clinicians and turn that interest into effort, the resources should have certain features, Pugno said. First, relevance to daily practice is crucial, he said, echoing comments made by previous speakers. There should also be easy access to the resources, such as a website that contains most of the information that will be helpful to clinicians and patients. Integration with electronic health record decision support can also be important, as can frequent updates aimed at addressing timely questions from patients. It is also effective to offer learning in small chunks at the convenience or time of need of the clinician. The Family Doctor website (FamilyDoctor.org) developed by AAFP has a number of these attributes.

"Challenges persist," Pugno concluded, and "there are no magic answers." But examples of things that work exist and can point the way toward future innovations and improvements.

MOTIVATING CLINICIANS

One of the biggest barriers to making genetics an integral part of continuing education for health professionals is motivating the clinicians to participate, which in turn depends upon them recognizing that there is a significant gap in knowledge that could impact their care of patients. As Raby described it, the clinicians need to "understand that there is demand for [this knowledge], that there is need for it." Scheuner pointed to the importance of convincing the leaders of an institution of the need for such education. For example, she talked with the leader of the Veterans Affairs Greater Los Angeles Healthcare System clinical service about BRCA1 and BRCA2 genetic testing, to which the leader was "very receptive." Clinicians also were interested in educational activities because they helped provide the context for the new tool they were using to obtain family histories.

Raby said that clinicians could also be motivated by the coming "tsunami" of readily available genetic information. The public is going to expect physicians to have an understanding of genetics, whether the genetic information they want explained comes from testing at a clinic or from personal genetic testing. Testing for research and clinical purposes has been increasing dramatically, and this testing is producing information that is directly relevant to patients, even if it is not yet actionable. With "most of the patients I see," Raby said, "the thing they are most grateful about is that I can actually tell them what is wrong with them, and I can explain to them why this has happened to them. The more information that is going to be out there, the more we are going to have to tell them."

Pugno said that if clinicians are to be interested in continuing medical education, they have to be able to answer "Yes" to three questions: Does this benefit my patients in a measurable way? Does this save me and my patient time? Does this save me and my patient's money? "If it addresses those three things, you will be beating them away from the door." David Davis summed up the issue of motivation in this way: "Knowing that every clinician gets up in the morning trying to do the best job they can, that is the clue to engagement for all of us."

EDUCATION AND QUALITY IMPROVEMENT

Education initiatives and quality improvement initiatives are typically conducted by two different groups of people within health care institutions. The critical question, Djuricich said, is how to integrate these two sets of activities at the point of care. Electronic health records may provide the answer, he said. These records cannot contain 45-minute educational modules, but they can contain what he called "educational nuggets." Packets of information that are quite short, less than a minute or two long, could help a clinician at the point of care, when that information is needed most to help a particular patient.

The accreditation of quality improvement projects for physicians and for multidisciplinary allied health staff teams has worked well for some institutions, according to one workshop participant. "For example," the participant said, "I work with pathologists, and they are constantly doing quality improvement with regard to turnaround time or improving laboratory reports." Physicians are extremely busy, but if they can receive continuing medical education in something that they are already doing, they can use that information to improve the quality of care for their patients.

Davis said that education will fail if the improvement component is not included. "If there are Venn diagrams, they overlap extensively." It is necessary to begin thinking about including a focus on improving the quality of patient care as a component of continuing education, he said.

Seibert pointed to the more general need to connect practice environments with educational communities in more formal ways. For example, many medical schools have adopted more integrated curricula. There are places where physicians, nurse practitioners, and other providers can work together on problems within the health care system, she said. "That may bring those educational pieces together."

4

Graduate Health Professional Education and Post-Graduate Training

Important Points Highlighted by the Individual Speakers

- Medical organizations that accredit residency and fellowship training or that certify physicians for independent practice can shape the content of graduate and post-graduate education programs.
- Standards or guidance for the clinical use of genomics findings is needed to take advantage of accreditation as a lever for change.
- Pharmacy organizations that accredit residency and fellowship training are updating their competency requirements in genetics and genomics for pharmacists to include training in applied pharmacogenomics and broader exposure to genomic medicine.
- Focusing on the critical points to master relating to a specific disease process or condition is effective for learner retention.
- The following core competencies are emphasized in post-graduate residency or fellowship training: patient care skills, clinical knowledge, practice-based learning and improvement, communication skills, professionalism, and systems-based practice.

HOW ACCREDITATION PROCESSES COULD BE USED TO PUSH GENETICS AND GENOMICS LEARNING

The Accreditation Council for Graduate Medical Education (ACGME), the body that accredits physician residency and fellowship programs, could serve as a lever for expanding genomics education at the graduate and post-graduate levels, said Kevin Weiss, ACGME's senior vice-president for patient safety and institutional review. The organization not only responds to changes in medical education but is one of the forces that shapes those changes.

The development of medical professionals is not a linear pipeline but a series of interrupted, challenging activities, Weiss said. This collection of activities includes pre-medical studies, medical school, specialty (residency) education, subspecialty (fellowship) training, and continuing education. Underlying these activities are the accrediting organizations and other bodies that certify individuals. "There is a series of tickets that have to be punched, both at the educational organization level, as well as the individual," Weiss said. "You can't just push one and expect the entire [collection of activities] to respond. You almost have to work as a group."

This is the context within which ACGME works. Its mission, Weiss said, is to "improve health care by assessing and advancing the quality of resident physicians' education through accreditation." In the United States, roughly 110,000 to 120,000 people are taking part in residency programs accredited by ACGME. These training programs must address six core competencies: patient care and technical skills; medical knowledge; practice-based learning and improvement; interpersonal and communication skills; professionalism; and systems-based practice. The chief focus of a residency program is on molding individual trainees into highly developed clinicians. However, Weiss said, "graduate medical education is not just about preparing clinicians for independent practice. It is also about preparing them to become proficient teachers and managers of resources, roles that take a lot of time and effort to develop."

ACGME is in the midst of a major effort to overhaul its accrediting process over the next 10 years, creating what is known as the Next Accreditation System (Nasca et al., 2012). The organization is shifting to focus continuously on improved patient outcomes. The current goal is to encourage program sponsors to maintain a humanistic educational environment that assures the safety and quality of the care that residents provide to their patients both today and in their future practices. Weiss also said that the organization may one day incorporate patient outcomes directly as a measure in the accreditation process. "We are at least a 5- to 7-year cycle before I think any of us can see that happening easily," he said, except perhaps with the clinical data registries that some health systems have been establishing.

To reach ACGME's current goal, the Next Accreditation System was built as a continuous accreditation model with annual updating based on data that residency programs must submit each year. This system includes a 10-year "self-study visit," Weiss said, and also a revision of standards every 10 years at a minimum, with reviews occurring much more frequently than that. To stay accredited, medical centers and institutions that sponsor residency or fellowship programs must also undergo a Clinical Learning Environment Review visit.

The annual review process for accreditation examines a series of outcomes, including the particularly important feature of milestone data, which is one of several elements used to assess residents' progress. The residents must demonstrate mastery of the requisite skills, knowledge, and behaviors in each of the six core competencies. "You can begin to see how competency domains may translate from genomics and genetics into a milestone," Weiss said. The milestones also create a framework of observable behaviors and attributes associated with residents' development as physicians.

Aggregate performance on the milestones will be used as 1 of 10 indicators for measuring a program's educational effectiveness as part of ACGME's continuous accreditation monitoring, Weiss said. The milestones "are probably even more important for residency programs because they can help guide curriculum development," he said. "They can help target individual residents for specific needs and improvement for the underperformers." Medical centers and institutions that sponsor residency or fellowship programs must undergo a learning environment review every 18 to 24 months, examining six focus areas: patient safety, supervision, professionalism, health care quality and health care disparities, duty hours and fatigue management, and transitions of care.

There is a great opportunity for ACGME to assimilate genomics education because "we can broadly shape the thinking in the residents and fellows through our requirements," Weiss concluded. As with the introduction of CT scans decades ago, the medical world is now facing "a rapid and not well-rationalized diffusion" of technology, he said—in this case, genomic technology. "We need to have a set of standards or some sort of guidance that is a clear signal [versus the] noise that really can take advantage of a lever that is as strong as accreditation."

A VIEW FROM THE PHARMACY ON IMPROVING GENETICS EDUCATION

The pharmacy profession has made substantial progress in the clinical implementation of pharmacogenomics and in national education efforts, said Grace Kuo, professor of clinical pharmacy and associate dean for academic clinical affairs at the University of California, San Diego. Pharmacogenomics promises to optimize drug therapy, minimize the trial-and-error approach to prescribing, and prevent avoidable adverse drug reactions. More than 100 medications approved by FDA include pharmacogenomics information on their labels at present,[1] and the number has been rising. Finally, applied pharmacogenomics offers a clear opportunity for spreading genomic medicine, as many institutions have adopted pharmacogenomic testing as an initial foray into genomic medicine practice.

More than half of the roughly 287,000 pharmacists practicing in the United States in 2013 did so in the community pharmacy setting, with another 22 percent practicing in hospital settings and 2 percent in non-patient care settings (e.g., administration, teaching, or research).[2] While post-graduate pharmacy residency and fellowship training is strongly supported by several national pharmacy organizations, relatively few Pharm.D. graduates pursue such training—largely because training slots are limited and post-graduate education is not a mandatory requirement for many pharmacy positions. In 2012–2013, only 42 percent of the 13,551 Pharm.D. recipients applied for residency, with half of those being accepted into accredited programs.[3] There is a critical need to continue developing the capacity for post-graduate clinical pharmacy training, Kuo said.

There is a growing profession-wide recognition that clinically applied pharmacogenomics offers an additional tool for optimizing the safe, effective, and affordable use of drugs, Kuo said. In 2009 the American Pharmacists Association convened a workshop to discuss the role of the pharmacist in using clinical pharmacogenetic data to direct patient care by effectively using electronic health records to exchange relevant health data among

---

[1] See http://www.fda.gov/drugs/scienceresearch/researchareas/pharmacogenetics/ucm083378.htm (accessed November 12, 2014).

[2] Data derived from the Bureau of Labor Statistics. See http://www.bls.gov/oes/current/oes291051.htm (accessed January 16, 2015).

[3] See https://www.natmatch.com/ashprmp/stats/2013applstats.html (accessed January 16, 2015).

the entire health care team (Reiss, 2011). Subsequently, large clinical implementation efforts were initiated to test the effectiveness of different care delivery models for pharmacogenomics (Hoffman et al., 2014; O'Donnell et al., 2014; Shuldiner et al., 2014), with various research studies, surveys, and focus groups underscoring the need for focused education in genetics, genomics, and pharmacogenomics (McCullough et al., 2011; Sansgiry and Kulkarni, 2003).

The PharmGenEd project was also developed as a continuing education curriculum focusing on pharmacogenomics primer concepts and clinical applications in various therapeutic areas (Kuo et al., 2011). Disseminated to health care professionals and students through a train-the-trainer program, the PharmGenEd project makes shared curriculum modules, video seminars, and relevant journal articles available online through its website (https://pharmacogenomics.ucsd.edu). "The program has now reached over 100 countries with 23,000 unique users and over 3,000 registrants," Kuo said, and it has been used by pharmacists, pharmacy technicians, pharmacy students, physicians, nurse practitioners, genetic counselors, researchers, and others.

PharmGenEd also evaluates outcomes and processes through two online modules, Kuo said. Continuing education credit can be earned through the American Society of Health Systems Pharmacists (ASHP) by passing the self-assessment exam on the society's website. For both modules, 97 percent of all users taking the exams passed.

While the pharmacy profession continues to play a key role in clinical pharmacogenomics, there is increasing momentum toward team-based care and interprofessional collaboration among front-line clinicians—both in genomics and in other clinical paradigms, Kuo said. To foster such collaboration the Genetics/Genomics Competency Center for Education (G2C2) project of the National Institutes of Health was instituted. The project provides high-quality educational resources in genetics and genomics for health care educators and practitioners.[4] G2C2 has so far collected educational competencies for nurses, physicians, physician assistants, and genetic counselors, and it is currently in the process of adding a new educational competency on genetics and genomics for pharmacists (Feero et al., 2012), which builds on a previous one developed in 2002 by the American Association of Colleges of Pharmacy (Johnson et al., 2002).

[4]See http://www.g-2-c-2.org/index.php (accessed November 13, 2014).

In order to adequately prepare and train the future pharmacy workforce to practice effectively, pharmacy faculty, national professional organizations, and the Accreditation Council for Pharmacy Education recently announced that the forthcoming 2016 accreditation standards for all U.S. pharmacy schools will, for the first time, require that pharmacogenomics be taught to all students. An important next step, Kuo said, may be to develop equivalent standards for post-graduate residency and fellowship training programs, although these are accredited by a separate organization (ASHP).

DRIVING PHYSICIAN LEARNING THROUGH THE BOARD CERTIFICATION EXAMS

Certifying boards define a discipline through certification, they track trends in medical knowledge application, and they provide assessment, which can also drive learning, said Rebecca Lipner, the senior vice president for evaluation, research, and development at the American Board of Internal Medicine (ABIM), Founded in 1936, the not-for-profit ABIM is an independent umbrella organization representing 24 certification boards in medical specialties, including internal medicine, allergy and immunology, medical genetics, preventive medicine, and urology. The organization's mission is "to enhance the quality of health care by certifying internists and subspecialists who demonstrate the knowledge, skills, and attitudes essential for excellent patient care." Roughly one in four practicing physicians in the United States is certified in internal medicine, which itself includes a multiplicity of subspecialties ranging from cardiovascular disease to transplant hepatology. Generally, the domains of medical knowledge or content that these disciplines cover are large, Lipner said, and genetics can fit into many of these areas.

To become board certified in internal medicine, a physician must complete a 3-year internal medicine residency with 2 to 3 years of further fellowship training required for subspecialty certification. Doctors must achieve satisfactory faculty ratings in the six core competencies specified by ACGME and pass a summative, high-stakes cognitive examination, Lipner said.

ABIM often uses board certification exams to help define the breadth of a medical discipline, Lipner explained. The advantage of a written exam is that "you can put a lot in there in a short period of time and cover a lot of area," but the trade-off is that "you usually do it in an artificial way, like with multiple choice questions or short answers." The idea is to broadly sample a physician's knowledge, judgment, attitudes, and ability to diagnose and treat patients, typically using questions involving clinical vignettes about patient cases. The test may require recognizing both common and rare conditions. Among the key features of these exams, Lipner said, are that they assess the core knowledge that a physician "should know and carry around in their head without looking it up" and that they emphasize content that reflects current best clinical practices, which, Lipner noted, "might be changing out from under us."

The starting point for ABIM's pro-active process to fit applicable content into the certification exam is a "practice analysis" or job analysis. This is a systematic procedure for collecting practice-related information—usually through a survey using a questionnaire developed out of research and input from focus groups and subject-matter experts—and defining the knowledge base in the discipline. The purpose of the practice analysis is to link the knowledge, skills, and attributes that are tested on the exam with the responsibilities of the profession. "Somehow, we have to create an exam, but we don't want to do it without understanding where the profession is coming from, what are the tasks that have to be done on that job," Lipner said. "Tasks are important, and person-oriented skills are also important."

Based on its analysis, Lipner explained, ABIM creates a blueprint to frame critically important exam characteristics (including the weight or percentage of the test devoted to different content areas), provide guidance to expert test question authors, and document historical performance. The blueprint for the medical oncology certification exam, for instance, shows that questions about genetics and tumor biology make up 3 percent of the exam. (That percentage does not, however, include many additional questions about genetic testing that are embedded within other content categories such as breast cancer and colon cancer.)

The evaluation of exam items takes into consideration two factors: difficulty, or how hard or easy a question is; and discrimination, or how well an individual question captures the overall ability of the examinees. These factors indicate whether certain content areas are obsolete and whether emerging medical evidence may be outpacing question content, Lipner said.

The exam blueprint is evaluated through an annual review process by subject-matter experts who assess changes in clinical practice guidelines and emerging content or practices to determine which topics should be included on the exam. The annual review is augmented by surveys or post-exam comments that offer feedback from doctors on, for instance, whether the blueprint is missing or overemphasizing certain content or skills.

Interestingly, cognitive science research (Larsen et al., 2008; Roediger and Butler, 2011) has found that "the act of taking an exam improved performance above and beyond simply studying for it," Lipner said. "Frequently taking tests is actually a good thing." Currently the board certification exams are required every 10 years. Because this "is probably not as good for retention [of information]," she said, ABIM is sponsoring an initiative called Assessment 2020 to explore the best practices for assessing physicians' skills that would be in addition to, or in place of the comprehensive 10-year exam.[5] Thus far the discussion has centered on such ideas as testing more frequently, targeting areas of weakness for individual physicians, and holding an open-book exam, Lipner said. Alexander Djuricich of Indiana University noted that it is possible to send personalized follow-up quizzes to individual examinees' smart phones. He also pointed out that the *New England Journal of Medicine* offers a learning program called Knowledge Plus™, which delivers test review questions for the internal medicine board exam via smartphone and tablet. The program uses adaptive learning technology to assess a learner's progress and identify content to reinforce with additional questions or content. ABIM is also planning to provide individual physicians more detailed feedback on areas for improvement relative to responses for specific content, Lipner said. The aggregate performance results will be shared with medical specialty societies in order to identify the areas of strength and weakness in the group of physicians completing residency training in order to assist quality improvement within education programs.

A VIEW FROM THE LEARNER'S SEAT

Offering insights from the learner's perspective, Samantha Bazan, a public health nurse in her second year of nurse practitioner training at the Uniformed Services University of the Health Sciences, shared her thoughts

[5]See http://assessment2020.abim.org (accessed November 12, 2014).

on effective teaching and learning in genetics. Bazan agreed with David Davis's earlier observation that simply providing more information in graduate and continuing education is not necessarily the best approach. "You can only retain so much information," she said. An alternate approach could include a sharp focus on the critical points to master relating to a specific disease process or condition—including genetic implications or associations. "Those things are going to be what sticks with you in clinical practice that triggers your brain to say, 'Oh, okay, I remember this. This is something we need to look for.'"

It is also valuable for trainees to understand why a genetic test is needed, what it is going to tell the patient, and what the implications of the results are, Bazan said. "If we learn that in our graduate programs, then we can better explain to the patient why we think they either need to go see a specialist or why we think, hey, you don't need this."

As far as preferred learning tools and strategies, Bazan said she finds UpToDate useful. During moments of downtime—for instance, while waiting to present a patient case to a preceptor—she will use the resource on her smartphone as a knowledge check. Clinical practice guidelines are also helpful for new practitioners, Bazan said, especially guidelines with algorithms that direct decision making. "Case studies are [another] important way to learn," she said.[6] "But it is more productive to work through case studies together in class, where you can benefit from hearing others' thought processes in reaching a conclusion, rather than being asked to do it on your own."

Bazan confessed to a "love–hate relationship" with the patient simulation center. "I absolutely hate it," she said. "I hate being videoed and having to go back and look at it." But the simulation sessions offer more realistic training than role-playing between students and faculty. "When you are actually doing it with simulation center patients who are there role-playing professionally, it works out really well. I have learned a great deal from that."

---

[6] Additional resources on case studies in genetics and genomics can be found at: http://g-3-c.org/en (accessed January 16, 2015).

DISSEMINATING GENETICS AND GENOMICS EDUCATION INTO PRACTICE

Graduate students can occasionally leap ahead of the faculty with their assimilation of knowledge of genetics and genomics, one workshop participant said. Three years ago the participant's institute started offering medical students the option of concurrently enrolling in a master's degree program in genomic medicine. The institution anticipates that medical school graduates with this additional degree will end up being an educational resource for the practices that they are entering.

Weiss said that "there is a real excitement to teaching the learners and getting them as an inoculum there to make change." However, as an isolated strategy, it might carry higher risk. When residents have received quality improvement training, but then move to practices that are unsupportive of this type of approach, the enthusiasm and engagement of that trainee can "just die on the vine," Weiss said. "That is a very painful thing to watch." Faculty need to plan for that potential circumstance, Weiss said, to keep young learners' engagement in quality improvement or genomics going. He also suggested that one strategy for spreading genomics knowledge into residency programs might be to encourage collaboration between faculty and residents on experiential projects.

Bruce Blumberg said that at Kaiser Permanente a similar issue arose related to efforts to provide medical residents with education about quality improvement. "We are learning pretty rapidly that we have faculty that don't know very much about quality improvement," he said. "Whenever a new body of knowledge comes into a field, you have that problem—where the learners and the people recently coming out of training have a skill or a body of knowledge that doesn't necessarily exist throughout the faculty."

Given such obstacles, implementing a genomics education component for currently practicing physicians is a grand challenge. One participant, for example, described how medical students at his institution could take an elective course in personalized medicine and genomes that offered them (and interested faculty) the chance to analyze either their own pharmacogenomics data or data from an anonymous person; then throughout their second year the students are taught how that genetic information could be useful in clinical practice. However, the faculty in charge are not entirely comfortable with the genetics material themselves, the participant said, and thus when students enter their third, fourth, and residency years, the follow-through on this earlier learning is lost.

Assessment is a chief driver of learning, and at times a new body of information must enter a field of practice. The critical challenge, Lipner said, is determining which persuasive sources to trust in terms of which new information is critical to assimilate.

5

Continuing Medical Education

Important Points Highlighted by the Individual Speakers

- A needs-driven, learner-centric, evidence-based, outcomes-oriented, and practice-embedded continuing medical education system can contribute to improved quality of care and patient outcomes.
- The accreditation system is designed to recognize and promote institutional and personal attributes that are effective in maintaining competencies, performance, and patient outcomes.
- Innovative methods in continuing medical education, such as simulations for workplace learning, massive online open classes, and Web-based learning portfolios, can help produce the changes that are needed as the effects of genetics on clinical practice continue to grow.
- Partnerships among professional societies can guide the development of educational initiatives and improve genetic literacy.

More than 2 million people each month interact with the continuing medical education system (ACCME, 2014). These interactions involve not just physicians but nurses, pharmacists, nurse practitioners, physician assistants, and other allied health professionals. While the current system has great potential for enhancing genetics knowledge among clinicians, doing so may require innovative educational approaches to ensure that practicing health professionals know how to apply emerging genomic information to clinical practice.

MAJOR FEATURES OF CONTINUING MEDICAL EDUCATION

Continuing education contributes to lifelong learning in any career, said Ann Karty, the medical director in the Continuing Medical Education Division at AAFP and chair of the Council of Medical Specialty Societies' Conjoint Committee on Continuing Education, which consists of continuing medical education directors from 25 member societies. Continuing medical education introduces learners to new technologies, innovative advances, and research. It serves to maintain, develop, and increase the knowledge, skills, and professional performance and relationships that a physician needs to provide services for patients, the public, or the profession. In this way, it contributes to improved quality of care and patient outcomes.

Credit is the "currency" that physicians and other health professionals earn by taking part in continuing education activities, Karty said. Physicians and health professionals use these credits to maintain licensure, specialty board certification, credentialing, membership in professional societies, and other professional privileges, with the requirements for earning a particular credit being determined by the organization responsible for certifying that credit.

As continuing medical education is currently set up, a curriculum is designed to lead teacher and learner to reach toward desired objectives. The components of a typical curriculum include aims and goals, content or subject matter, experiences, and evaluation, Karty said. The curriculum, when implemented, results in new knowledge, skills, and attitudes or abilities, sometimes known collectively as KSAs. Ideally, the curriculum should be needs-driven, learner-centric, evidence-based, outcomes-oriented, and practice-embedded. It should be designed to bridge the gap between actual practice and ideal practice. Box 5-1 describes the perspectives of various specialists on education about and the use of genetics and genomics; these perspectives point to some needs within the community that genetics education could fill.

BOX 5-1
Perspectives on Genetics Education

Before the workshop, Ann Karty asked members of the Council of Medical Specialty Societies to comment on educational requirements in genetics and genomics. Here are some of the responses she received:

- "When we've done programming on genomics, it's been foundational, didactic activities because it's primarily new information that our members aren't familiar with. And because much of it in our field is new, there's not much clinical application yet. For that reason, the whole issue of just-in-time isn't relevant for us now because we don't have immediate clinical applications yet."
- "From a general sense, we make more use of point-of-care tools. We haven't made point-of-care education part of our program, and our certifying board is big on providing tools as part of the MOC [maintenance of certification] self-assessment/performance improvement process."
- "We're in the process of analyzing our annual meeting programming process, and one of the issues we're looking at it is strengthening the basic science education we offer. I think if this gains traction, we might start seeing more."
- "Genomic data cannot be used to improve health if it cannot be accurately communicated and correctly understood. . . . [There is] tremendous variability in the representation of genetic test results [and in] the efficient exchange and use of this information by clinicians and researchers."
- "Credit could be an incentive, but point of care tends to be such small amounts that tracking and claiming might actually be a barrier."
- "How can we incorporate genomics into clinical decision support systems [even as it relates to] the goal of pharmacogenomics . . . and using the right drug, the right dose, at the right time."
- "[Our] mission . . . is to facilitate communication, collaboration, training, and networking for researchers working at the interfaces between biomolecular and clinical data, . . . to advance the clinical use of genomics data and the fields of genomics and translational bioinformatics, . . . furthering the practice of precision medicine. Our members are highly involved in the development of software tools, methods, and standards that are necessary for the clinical application and

interoperability of genomics data, [and] our members also participate in the development, integration, and delivery of knowledge content to clinicians."

- "Due to both rapid changes in knowledge and the size of the genomics domain, we strongly believe that it will be necessary to provide context-dependent, just-in-time genomics education to clinicians (e.g., through clinical decision support systems, including alerts and curated knowledge bases). Given the complexity of genome biology, however, we believe clinicians must also have a solid base of knowledge about genomics in order to effectively understand and utilize the just-in-time material."
- "While there may be some instances where offering continuing medical education credit for informatics-based courses would have direct impact on patient care, there are probably many more opportunities to offer continuing medical education credit for courses that provide general knowledge about genomics or specialized (gene- or phenotype-specific) interactions. Therefore, the majority of continuing medical education credit should likely be directed toward content rather than technology."
- "Actual application to clinical practice will require a culture change as well as a technological one. Some clinicians don't want to use genetic data in their practice until they have randomized clinical trial (RCT) evidence that it helps. However, it is not financially or statistically possible to perform an RCT for every genetic variant and every possible clinical application of that variant. The application of genetic data to clinical practice will not grow significantly until clinicians are both able and willing to apply the interpretations from certified genetic counselors in the absence of RCTs when it might help improve outcomes."
- "[We] don't have any technology-based point-of-care tools or point-of-care education right now, but have produced pocket cards and 'analog' tools in the past. . . . It is possible that [a] registry could eventually track and provide links to (yet to be created) educational activities on genetics, but we are a year-plus away from this capability."

"These comments may be potential pieces to a needs assessment to help create more continuing medical products," Karty observed.

The content of continuing medical information depends on the knowledge, skills, or attitudes that need to be transmitted to the learner. The way in which one measures outcomes depends on the best format (e.g., live activities, self-study, journal reading, manuscript review for journals, point-of-care learning, online learning, performance improvement in practice, and translation to practice) for addressing the gap that has been identified and also on the learning objectives that have been designed to address that gap. Karty described a framework for the assessment of continuing medical education that ranges from "does" (performance) to "shows how" (competence), "knows how" (procedural knowledge), and "knows" (declarative knowledge) and that ensures that education is free from commercial conflicts of interest.

Requirements for continuing education can vary greatly between states, Karty said. Requirements also can be highly specific to subspecialties. For example, the American Board of Medical Specialties has 24 specialty boards and more than 145 specialties and subspecialties. Despite this variation, however, particular themes appear across requirements, including patient care, interpersonal and communication skill, professionalism, medical knowledge, practice-based learning, and systems-based learning. Within these themes, AAFP includes genetic and genomic content such as family history taking, red flags, cultural competency, evidence-based medicine, health information technology, and specific content related to certain cancers, cardiovascular disease, congenital hearing loss, developmental delays, and ethical, legal, and social issues.

Finally, Karty said, not all practitioners need to know the same thing, despite continuing medical education requirements that force all clinicians to learn about subjects that they may or may not use. A better approach might be individualized learning portfolios for practitioners, which reveal both what an individual practitioner knows and what he or she does not know. In this way, a testing mechanism could point out what a particular individual needs to learn. "Mandatory continuing medical education is not what I would necessarily move toward," Karty said. "I would love to have individualized portfolios telling me where my deficiencies are so that I could educate myself."

EVIDENCE-BASED CONTINUING MEDICAL EDUCATION

With appropriate formats, continuing medical education can create desirable physician attributes and competencies and contribute to organizational self-assessment and improvement, said Murray Kopelow of the Accreditation Council for Continuing Medical Education (ACCME). Continuing medical education can be based on evidence-based requirements with the expected results being changes in competence, performance, or patient outcomes. It also can be done in a way that manages boundary issues with commercial interests.

As an example of the potential impact of continuing medical education, Kopelow mentioned work that he had done several years earlier in the White House Office of National Drug Control Policy on the misuse of prescription drugs. A continuing medical education program on mitigating risks for these products was developed that resulted in about 2,000 activities that reached about 60,000 participants. Such success stories could be repeated elsewhere, Kopelow said.

Under ACCME, the accreditation system is based on evidence of what is effective for changing competencies, performance, and patient outcomes, Kopelow said. It encourages the right formats for particular educational needs. A recent ACCME report identified 39 systematic reviews that together describe an evidence-based approach to designing continuing medical education (Cervero and Gaines, 2014). The report found that continuing medical education improves both physician performance and patient health outcomes, although the more reliable effect was on physician performance, Kopelow said. It also determined that continuing medical education is more effective when interventions are interactive, longer, and reinforced.

Kopelow particularly emphasized the importance of predisposing learners to seek out the information they need, because learners often are not aware of how much they are missing or what they cannot do. Most learners are committed to excellence and to lifelong learning, but they do not necessarily know what they do not know.

Kopelow also noted that accreditation requires that continuing medical education providers involve undergraduate medical students and graduate medical students in the planning, delivery, and evaluation of continuing medical education. Providers that are not medical schools or residency training programs have been pushing back on this requirement, saying that they do not know how to involve students in the education they offer. But, Kopelow said, such organizations have many ways of engaging students in their work.

Since 2007 ACCME has also had a program that covers interprofessional education, which is particularly important in genetics. Nursing, pharmacy, and medical institutions can be accredited through one process to do "education for the team, by the team," Kopelow said. The accrediting organization also is changing its criteria for accreditation with commendation to encourage interprofessional practice.

Kopelow cautioned that many intervening variables lie between continuing medical education interventions and changing the health of the people of the United States. In general, the greater the number of intervening variables there are, the greater the likelihood is that there will be no change as a result of continuing medical education.

Furthermore, Kopelow observed, continuing medical education is not necessarily always designed to change behavior. It may be designed to give people ideas about what they can do or how they can overcome barriers. "Don't be constrained by any current structure that you observe in continuing medical education," he said. "This can be all reimagined and reinvented. The formats of education are only the formats that people have thought of so far." Even failed experiments in continuing medical education can be useful in revealing something that did not work, he said.

Innovative technologies are beginning to be used in health professional education, Kopelow said. (See Chapter 3 for further details.) Young physicians tend to learn through a variety of new methods, including group activities such as interactions on social media. "What people need to know is not how to plan a Twitter interaction but how to judge the information," he said. "When they hear it, they need the tools to judge what they are hearing. How does it fit with the literature that is available? Where do they go for wisdom and judgment about how to apply it in practice?" This educational trend will likely continue to grow. The physicians who will be entering practice in 10 years are now starting college, and those who will be entering practice in 20 years are in the fourth grade, Kopelow observed. "Those are the people that we need to plan for."

DEVELOPING COMPETENCIES THROUGH CONTINUING MEDICAL EDUCATION

There is a distinction between continuing education and continuing professional development, said Constance Goldgar, associate director of the University of Utah Physician Assistant Program. The former refers to educational activities that serve to maintain, develop, or increase knowledge, skills, and professional performance, while the latter is a multifaceted approach to the acquisition and application of knowledge during the practice life of a health care professional. In particular, continuing professional development is learner-driven, tailoring instruction to individual needs, and it encompasses lifelong learning across the span of one's career (IOM, 2010).

Educational activities can be both systems-based and competency-based, with the former relying on incentives and programs and the latter involving individual-directed learning. A particularly useful framework, Goldgar said, involves competencies, which she described as the observable abilities of health professionals. Such a framework defines the outcomes desired at the end of an educational pathway.

These competencies are based on a foundation of evidence, ranging from expert opinion to systematic reviews to computerized decision support. This evidence needs to be clinically meaningful, implementable, and available at the point of care, Goldgar said. The ability to modify competencies, which are different today than they would have been 20 years ago, is essential to adaptive change.

Changes in the genetics education environment are being driven by a number of forces, including research, technologies, regulations (such as those governing MOC), funding, quality improvement, health disparities, changes in health care institutions, the genetics workforce, and the clinician population. Meanwhile, broader social, technological, economic, environmental, and political changes are occurring that influence these driving forces. Identifying the driving forces can reveal uncertainties or polarizing forces that lie ahead, Goldgar said.

Using innovative methods in continuing professional development can help bring about the changes that are needed as genetics increases its impact on clinical practice. Among the interesting innovations that have been discussed are advanced simulations for workplace learning, audits of patient populations using electronic health records, massive open online classes, collective intelligence through networking, and Web-based learning portfolios. "The future is wide open," Goldgar said.

PROFESSIONAL SOCIETIES IN GENETICS EDUCATION

The amount of data in a genome sequence is enormous, said Michael Murray, an internist and geneticist at the Geisinger Health System in Pennsylvania and a co-chair of the Inter-Society Coordinating Committee for Practitioner Education in Genomics (ISCC) for the National

Human Genome Research Institute. Figuring out how to interpret the data will be a 100-year project.

In a 2012 survey of more than 480 internists conducted by the American College of Physicians, half of the internists reported that their volume of genetic testing had increased over the preceding 2 years, while 13 percent reported that they had not ordered a genetic test in the prior 2 years. The overwhelming majority—97 percent—said that they were willing to devote time to continuing medical education in genetics. Although this suggests a "ready and willing" audience, Murray said, as discussed previously, education must be relevant to the needs of clinicians in order to retain their interest. If it is not relevant, they will drift away, and it will take time and effort to get them back. "We have to use our opportunities wisely," he said. "In that effort, we have to engage the people who are doing this work and work with them as we develop continuing medical education opportunities, not just talk to people like me who are genetics professionals."

Launched in 2013, the purpose of ISCC is to facilitate interactions among medical professional societies and the institutes and centers of the National Institutes of Health in order to exchange practices and resources related to genomics education and clinical care. Its goals are to

- Promote the sharing of educational approaches and the joint identification of educational needs in order to improve genomic literacy and to enhance the practice of genomic medicine.
- Jointly identify the needs of professional societies and clinicians in filling in gaps in evidence and knowledge and in providing effective educational efforts.
- Offer partnership and available expertise to member societies to guide the development of educational initiatives.

ISCC has divided its activities into four areas: competencies, educational products, engagement of specialty boards, and use cases (Manolio and Murray, 2014). The group reviews existing competencies in genomic medicine education as well as guidelines for the use of genomics, examines surveys and other sources to see what competencies fit into current practice, and interacts with individual professional societies to determine how amenable they are for developing competences in genetics for their profession (Korf et al., 2014). ISCC also collects existing educational products from current members, identifies relevant federally funded resources and initiatives that could assist genomics education efforts and clinical practice, works to identify areas of emphasis for educational products (such as the ordering of genetic tests, counseling, or the return of results), and identifies new advances that may require educational initiatives. A major effort within ISCC is to determine the extent to which specialty boards have integrated genomics into their examinations and to engage directly with those boards that may not be doing so at this time, Murray said. The group also acts to link specialty boards with relevant professional societies that are already implementing genomics education or are looking to do so. Finally, ISCC collects existing use cases or develops general and society-specific use cases in genetics across five general topic areas: pharmacogenomics, family history, rare single-gene disorders, common diseases with genetic components, and whole genome/exome sequencing, including incidental findings. These use cases can be identified by specialty end users or based on existing competencies.

As Murray pointed out, competencies differ from entrustable professional activities (EPAs), although they may be closely related. While EPAs cannot serve as an alternative to competencies, they do represent a means to translate competencies into clinical practice. Five EPAs have been developed by ISCC so far:

- Family history EPA: Elicit, document, and act on relevant family history information pertinent to the patient's clinical status.
- Genomic testing EPA: Use genomic testing appropriately to guide patient management.
- Patient treatment based on genomic results EPA: Use genomic information to make treatment decisions.
- Somatic genomics EPA: Use genomic information to guide the diagnosis and management of cancer and other disorders involving somatic genetic changes.
- Microbial genomic information EPA: Use genomic tests that identify microbial contributors to human health and disease, as well as genomic tests that guide therapeutics in infectious diseases.

These EPAs are not the final word but rather are a framework for the development of competencies, Murray said. "The hope is that each group would take these, personalize them, and use them in effective ways within their specialty or their professional area."

Murray also affirmed the utility of the G2C2 resource (the Genetics/Genomics Competency Center for Education) referred to in Chapter 4.

Mapping educational needs to these resources reveals the holes in educational products, Murray said. "When we think about where we are and where we need to get to, this will allow individual practitioners or groups to . . . begin to know what we don't know, to realize where the holes are and where the good educational products are."

Finally, Murray briefly described ongoing work at Geisinger using whole-exome sequences on 100,000 patients in the organization's biobank. As part of the educational products associated with the project, physicians can sign up along with their patients to receive 5 to 10 whole-genome sequence reports. Through a mixture of lectures and self-study, providers learn about the genome report and then have a resource center from which they can access information. In the first round of sequencing results, more than 1,000 physicians and 800 advanced practitioners are delivering results to patients. In this setting, education is shifted to the moment when the provider is most likely to be interested in learning this information, Murray said.

6

Next Steps to Achieve Effective Genetics Education for Health Professionals

Education is a key component of translating genomics into health care, said Geoffrey Ginsburg, co-chair of the Roundtable. Workshop co-chair Joan Scott added that practical approaches are needed to make significant changes. In the final session of the workshop, a panel of previous presenters discussed, along with the other workshop participants, what needs to happen for health professionals to be effectively educated in genetics. First, they addressed the question of what needs to change. Then they explored how those changes can be realized.

WHAT CHANGES NEED TO BE MADE?

The number of people whose health care is influenced by genetics science or else the number of professionals who have integrated the issues of genome science into their practice needs to change, said Murray Kopelow of ACCME. (Box 6-1 lists the various changes suggested by workshop participants.) Genomics education must be relevant to a health professionals practice and should have an ultimate goal of improving patient outcomes, said Diane Seibert of the Uniformed Services University of the Health Sciences. Bruce Blumberg of Kaiser Permanente agreed with the need to build the case for the relevance of genomic information for practitioners, but he cautioned against overpromising. "Any new technology suffers from an exaggerated early prediction of impact and an underestimate of long-term impact," he said.

Competencies, once they are developed, provide a roadmap for providers and for the developers of educational resources, said Constance Goldgar of the University of Utah. Clinicians need to be reached in

BOX 6-1
What Changes Need to Be Made?

- Improved patient outcomes (Seibert)
- Genomics education for all health professionals (Seibert)
- The competencies expected of health care providers (Goldgar)
- The clinical care gap (Davis)
- Conflict-of-interest policies (Davis)
- Interprofessional education (Kuo)
- Awareness of genetic factors in health care or the number of professionals who have integrated genome science into their practice (Kopelow)
- The sense of urgency surrounding the issue (Weiss)

practical ways that do not require significant changes to their daily workflow, as this may garner resistance, she said.

The huge amount of information provided by genetics science will require thinking about education differently, a workshop participant added. New tools are going to be needed to explore the genome of each patient, including curated databases of genomic interpretations. But only some physicians, such as pathologists or medical geneticists, will need to know how to use all those tools. Other providers will need what the participant called "genetic sensitivity or curiosity." "I think moving the needle is just raising genetic awareness across all physicians," she said.

Some professions need and want more information than others, said Grace Kuo of the University of California, San Diego. For example, pharmacists are very detail oriented and want as much information as possible, while many physicians would consider that same amount of information to be excessive. Expert input will be needed to establish guidelines in these areas.

There is little sense of urgency that change has to occur, said Kevin Weiss of ACGME. To get something done within 3 to 5 years will require that genetics education be given a higher priority than is the case today.

There is a significant gap between what is known and what happens in the clinic, said David Davis of AAMC. "Here we are with knowledge," he said. "How do we translate that into practice? It is a huge field for us." He also emphasized the importance of conflict-of-interest issues. Standards for commercial support exist, he said, but are they enough in the current environment? "I don't know that we have an answer for that." Kuo also identified IPE as a significant area in which the needle needs to be moved.

HOW CAN THOSE CHANGES BE MADE?

The best way to make the changes that have been identified is to make genetics more relevant to clinicians, Seibert reiterated. (See Box 6-2 for a list of suggested ways to change genetics education in medicine.) For example, she said, the work being done at Geisinger (see Chapter 5) to provide genetics information when it is needed and in a real-world context is particularly promising. "I don't think there is a better way to gather information, absorb it and keep it than to present it in" a clinical context, she said. She also described the effectiveness of bringing in patients with genetic disorders to teach students about the issues involved:.

BOX 6-2
How Can Those Changes Be Made?

- Make genetics more relevant to clinicians (Blumberg, Goldgar, Kopelow, Murray, Seibert, Weiss)
- Encourage the taking of family histories (Goldgar)
- Target promising areas for the identification of genetic risk (Kuo, Seibert, Weiss)
- Identify a specific application that could help close the practice gap (Kuo, Murray, Seibert, Weiss)
- Increase and support interprofessional education (Goldgar, Johnson, Kopelow, Pugno, Raby, Scheuner)
- Explore disruptive technologies such as inexpensive sequencing and pharmacogenomics (Seibert)
- Include educational material and resources in laboratory reports (Raby)
- Work with accrediting and certifying bodies to improve and expand genetics education (Blumberg, Johnson)
- Adopt educational approaches that are more likely to change behaviors (Goldgar, Karty)
- Encourage self-assessments or practice reviews to identify provider needs (Kopelow)
- Develop champions for genetics education (Goldgar)
- Integrate education into the practice environment (Kuo, Murray)

"Educational interactions that involve an interested group of consumers . . . capture an audience and make it real." Providing lessons drawn from patient cases requires a cadre of peer reviewers to distill information with practice-based relevance, Kuo added. "It takes a lot of work, [but] we need to continue to do that."

The success of any educational effort in genetics will ultimately lie in the ability to engage clinicians and to make genetics and genomics interesting, said Michael Murray of Geisinger Health System. Work must be done on the backend to ensure that individuals come away feeling energized by and engaged in the encounter, Seibert said. Weiss suggested preparing an "elevator speech" for the chief executive officer of a major health system on why a particular genetic evaluation should be done for all relevant patients. "That exercise sharpens the question of what you could deliver and why you could deliver it," he said.

Goldgar pointed to the value of having students do pedigrees on their own families, which "engages them immediately." By working on their own family histories, they learn what a family history looks like in the clinic, where Goldgar also requires her students to do pedigrees. However, the challenge is that this activity is not typically reimbursed and thus once these students reach the clinic, they "unlearn everything because they say no one has time for this," she said. They see the value of taking the pedigree as students, but when they see that their preceptors forego this completely, they lose all the momentum that they had developed. "There is this whole disconnect between some of the basic things that we need students to be doing, and then unlearning them because you have 12 minutes to see a patient," she said.

The concept of a family physician needs to be revisited and rejuvenated, Benjamin Raby of Harvard Medical School and UpToDate said. Family physicians have traditionally taken care of families and thus have had a better understanding of the history and potential disease risks of a family's members. "I don't know how much that occurs today," he said. "Things are segmented out. The internists take care of the adults, and pediatricians take care of the children." A better approach, he said, would be to consider not just single patients but their families and whether genetic testing is worthwhile for them. "From a public health perspective and from the type of engagement that we want to have with our patients, one of the ways to raise the bar and show people how important genetics can be is the role that you can have not only for the patient sitting in front of you but for the patient's [family]." Obstacles exist, though, for this type of approach, Raby said. For example, testing for the purposes of counseling other family members is not usually covered by insurance companies.

The field needs to close the practice gap, Weiss said. From a systems or health care delivery perspective, he said, we need to ask: Of everything that is currently known in genetics and genomics, what one thing do we need to integrate into practice that would improve or save lives? Murray made a similar point when he described the need to identify a specific application that could help close the practice gap. This, Murray said, would be "the application where every provider in America says, 'You know, I really ought to be doing this for my patients,' or where patients come to me and say, 'Why aren't we doing that?'" For example, early data suggest that 2 percent of people may have a cancer predisposition gene or a cardiovascular disease predisposition. "The cost may become low enough and the importance of identifying that 2 percent might be high enough that we might reach the point where everybody says that we need to start knowing more about this and applying it to broad populations," he said.

Several areas of genetic screening have become well established in practice, such as newborn screening and prenatal diagnosis, as one participant noted. He urged that the field evolve "to the point that we could be doing adult screening toward the objective of identifying risk before pathology and intervening on the risk and the outcomes in the pathologic state." To convince the public that such screening would have utility, he suggested targeting a few areas where the identification of genetic risk would create an opportunity for intervention. "Two big killers in the United States are cancer and cardiovascular disease," he noted. "If I were to identify a project area that we could concentrate on and try to prove the principle that adult screening works to reduce morbidity and improve outcomes, I would say it is the cardiovascular area." Such a program could identify not just those at risk of heart diseases, such as people with hypercholesterolemias from receptor defects, but perhaps also those who might not benefit from receiving an intervention such as an implanted defibrillator.

Seibert suggested focusing on a medical issue like obesity, with its interactions between a person's individual genome and biogenome. "Patients are motivated about that," she said. "You might get more traction." Kuo suggested that pharmacogenomics testing might be a high-leverage application in terms of the potential to identify patients who will benefit most or have the least toxicity from a particular drug.

The value of negative results should not be overlooked, several participants observed. For example, Weiss pointed to the value of a genetic test that would predict if a person has a very low probability of colon cancer. If such a test eliminated half of the population's needs to do bowel preps and colonoscopies, he said, "you would be the star of the moment."

Several speakers spoke of the need for IPE. As Goldgar pointed out, within health care reform there is an increasing emphasis on team-based care, which opens a new door for IPE. Such an approach to education would get people out of their silos and would have particular relevance for genetics, where different practitioners will have different roles but will still need to work together. All practitioners, Kopelow suggested, can ask their institutions what they are doing in the area of IPE and whether they are working toward a jointly accredited system featuring seamless interprofessional planning and development. Today, individuals are accountable to their licensing and regulatory authorities, but teams of professionals also can be held accountable. "That is a new concept, to stand and fall together," he said.

Continuing medical education should be "a team effort for the team and by the team," emphasized Sam Johnson of Kaiser Permanente. Johnson also noted the existence of a paradigm shift in delivering continuing medical education, with a much greater emphasis on what, how, and to whom to communicate than on the specific venue. Finally, Johnson said that interprofessional teamwork has to dovetail with more than just education. Licensure, accreditation, and, most important, leadership are all involved.

The role of accrediting and certifying bodies in improving the health care system was also emphasized by Blumberg. These entities provide a link between education and quality outcomes, he said. Assessment can drive learning and create opportunities to be proactive in advancing the knowledge required of practitioners and organizations.

On a similar note, a participant said that a major driver of change for genetics education in medical schools was the inclusion of more genetic questions on the United States Medical Licensing Exam, with a genetics subscore being reported. When the students of an institution do poorly on the genetics part of a licensing exam, the dean of that institution pays attention, he said. One area in which the needle needs to be moved now is the application of genetics knowledge in practice. Such a change could be facilitated through the inclusion of genetics questions on specialty residency exams and participation of geneticists on item-writing committees for the various national boards.

Perry Pugno of AAFP pointed to the value of an annual conference on practice improvement that includes everyone on a health care team, "not just the physicians and nurses, but the receptionist and the medical assistant and the practice manager and the biller and everybody involved in that patient's encounter." In a similar vein, Raby reiterated the value of creating incentives to have people within a hospital shadow the health care providers in other clinics "because that is one of the best forms of education—to actually see what other people are doing." As an example of such an effort, Maren Scheuner of the U.S. Department of Veterans Affairs pointed to a program in which 150 internal medicine residents at the Veterans Health Administration hospital have 1 week in which they rotate through different subspecialty clinics. "It is the first time since I have been there in 5 years where I am able to interact with internal medicine residents," she said.

Disruptive technologies could foment major changes in genetics education, several workshop participants pointed out. Seibert noted that very cheap whole-genome sequencing could "change the world faster than we can keep up with it." Murray added that there is also the potential that if the field does not generate a plan and carry it out itself, others will dictate to the field how to do so instead. "My hope is that we will stay ahead of that curve," he said.

The potential of laboratory reports to serve as a portal for genetic education was a prominent topic of discussion. For example, Raby pointed out that including education within laboratory reports is a way of educating practitioners without forcing them to engage in an educational program. Furthermore, in this way geneticists can empower other subspecialties to act on genetic information.

However, Blumberg pointed out that information for clinicians on what to do next in laboratory reports assumes a perfect genotype–phenotype correlation and complete laboratory access to clinical information, but neither condition applies in the real world. "There are a lot of things about the indication for the test and the family structure and the patients' philosophy of care, etc., that the laboratory could never know," he said. "I would argue that it really is a dialogue between clinicians and laboratorians that ought to determine what next steps ought to be. It would be the rare, rather than the common, circumstance that the laboratory would know enough about the case and enough about the genotype–phenotype correlation to give advice as to what the next step ought to be." Kuo added that a lot of hospitals and clinics do not have laboratories that provide genetic testing at this time. "That is a real challenge in the practice sense."

Raby countered that a laboratory does not need the perfect genotype–phenotype correlations and the entire clinical picture to be able to make suggestions. For example, at the end of each UpToDate topic, authors are encouraged to make recommendations for action if appropriate. In addition, standards could be developed for deciding when information should and should not be included. Raby also pointed out that tremendous amounts of information are soon going to be available from multiplex testing and massively parallel sequencing. "The clinician is going to be faced not only with the result from the gene that they asked for, but from maybe 100 or more genes that they didn't ask for," he said. The result will be much greater costs and effort by clinicians, and patient and family anxiety could be increased.

The approach taken in delivering genetics education can make a big difference, several speakers noted. Medical education could be required to add components that are more likely to change behaviors, such as an interactive component, Goldgar suggested. Easier access to patient outcomes would help in the evaluation and design of continuing medical education, said Ann Karty of the Council of Medical Specialty Societies. Genetic counselors may also be well poised to be educators, observed a participant, because they are involved with many different disciplines.

Kopelow argued that the most productive use of a limited amount of interest by clinicians in genetics education would be to have them do a self-assessment or practice review to identify what their needs are. Even a short period of time devoted to this end could launch people on "a lifelong journey to try to solve their problems," he said. He also suggested that physicians could be taught to ask a few key questions that can open the door to the use of genetic information in a clinical encounter. "In the substance abuse world, they have this thing called screening and brief intervention," he said. "If you get the docs to ask [their patients], 'How many times in the last month have you had five drinks in one day?' that is all you have to do to open this whole world and do an intervention."

Finally, Goldgar pointed out that one way to move the needle is to develop champions across specialty areas for genetics education. Encouraging and supporting champions for change creates a "top-down, bottom-up approach" that can make a difference, she said.

CONCLUDING REMARKS

Co-chair Joan Scott concluded the workshop by thanking the participants and pointing to a theme underlying much of the discussions. "I was surprised and gratified to hear that there did seem to be a general agreement about what [needs to change]—that is the awareness within the broader community about the relevance of genomics."

References

ACCME (Accreditation Council for Continuing Medical Education). 2014. *Accreditation Council for Continuing Medical Education 2013 annual report: Executive summary*. Washington, DC: ACCME.

Al-Azri, H., and S. Ratnapalan. 2014. Problem-based learning in continuing medical education: Review of randomized controlled trials. *Canadian Family Physician* 60(2):157–165.

Bird, G. C., K. Marian, and B. Bagley. 2013. Effect of a performance improvement CME activity on management of patients with diabetes. *Journal of Continuing Education in the Health Professions* 33(3):155–163.

Bloom, B. S. 2005. Effects of continuing medical education on improving physician clinical care and patient health: A review of systematic reviews. *International Journal of Technology Assessment in Health Care* 21(3):380–385.

Cain, D. M., G. Loewenstein, and D. A. Moore. 2005. The dirt on coming clean: Perverse effects of disclosing conflicts of interest. *Journal of Legal Studies* 34(1):1–25.

Cervero, R. M., and J. K. Gaines. 2014. *Effectiveness of continuing medical education: Updated synthesis of systematic reviews.* Chicago: Accreditation Council for Continuing Medical Education.

Chaudhry, A., L. M. Glode, M. Gillman, and R. S. Miller. 2012. Trends in Twitter use by physicians at the American Society of Clinical Oncology annual meeting, 2010 and 2011. *Journal of Oncology Practice* 8(3):173–178.

Colbert, J. A., and D. A. Chokshi. 2014. Technology in medical education—Osler meets Watson. *Journal of General Internal Medicine.* http://link.springer.com/article/10.1007%2Fs11606-014-2975-x (accessed November 18, 2014).

Crain, E. F. 2014. Pediatrics and the lure of technology. *Academic Pediatrics* 14(1):4–7.

Davis, D., and R. Galbraith. 2009. Continuing medical education effect on practice performance: Effectiveness of continuing medical education:

American College of Chest Physicians evidence-based educational guidelines. *Chest* 135(3 Suppl):42s–48s.

Davis, D., M. A. O'Brien, N. Freemantle, F. M. Wolf, P. Mazmanian, and A. Taylor-Vaisey. 1999. Impact of formal continuing medical education: Do conferences, workshops, rounds, and other traditional continuing education activities change physician behavior or health care outcomes? *Journal of the American Medical Association* 282(9):867–874.

Davis, D., M. Evans, A. Jadad, L. Perrier, D. Rath, D. Ryan, G. Sibbald, S. Straus, S. Rappolt, M. Wowk, and M. Zwarenstein. 2003. The case for knowledge translation: Shortening the journey from evidence to effect. *BMJ* 327(7405):33–35.

Davis, D. A., J. Prescott, C. M. Fordis, Jr., S. B. Greenberg, C. M. Dewey, T. Brigham, S. A. Lieberman, R. W. Rockhold, S. J. Lieff, and T. E. Tenner, Jr. 2011. Rethinking CME: An imperative for academic medicine and faculty development. *Academic Medicine* 86(4):468–473.

Desai, T., A. Shariff, A. Shariff, M. Kats, X. Fang, C. Christiano, and M. Ferris. 2012. Tweeting the meeting: An in-depth analysis of Twitter activity at Kidney Week 2011. *PLoS ONE* 7(7):e40253.

Feero, W. G., G. M. Kuo, J. F. Jenkins, and M. A. Rackover. 2012. Pharmacist education in the era of genomic medicine. *Journal of the American Pharmacists Association* 52(5):e113–e121.

First, L. R., A. R. Kemper, K. Larson, and J. Puskarz. 2014. The gateway to journal improvements. *Pediatrics* 133(1):129–130.

Forest, C. P. 2012. The effect of audience response systems on adult learning: Evidence-based rationale and audience response systems implementation guide. *Journal of Physician Assistant Education* 23(4):54–59.

Forsetlund, L., A. Bjorndal, A. Rashidian, G. Jamtvedt, M. A. O'Brien, F. Wolf, D. Davis, J. Odgaard-Jensen, and A. D. Oxman. 2009. Continuing education meetings and workshops: Effects on professional practice and health care outcomes. *Cochrane Database of Systematic Reviews* (2):Cd003030.

Gaglani, S. M., and E. J. Topol. 2014. iMedEd: The role of mobile health technologies in medical education. *Academic Medicine* 89(9):1207–1209.

Grimshaw, J. M., L. Shirran, R. Thomas, G. Mowatt, C. Fraser, L. Bero, R. Grilli, E. Harvey, A. Oxman, and M. A. O'Brien. 2001. Changing provider behavior: An overview of systematic reviews of interventions. *Medical Care* 39(8 Suppl 2):II2–II45.

Hoffman, J. M., C. E. Haidar, M. R. Wilkinson, K. R. Crews, D. K. Baker, N. M. Kornegay, W. Yang, C. H. Pui, U. M. Reiss, A. H. Gaur, S. C. Howard, W. E. Evans, U. Broeckel, and M. V. Relling. 2014. PG4KDS: A model for the clinical implementation of pre-emptive pharmacogenetics. *American Journal of Medical Genetics, Part C: Seminars in Medical Genetics* 166c(1):45–55.

Interprofessional Education Collaborative Expert Panel. 2011. *Core competencies for interprofessional collaborative practice: Report of an expert panel.* Washington, DC: Interprofessional Education Collaborative.

IOM (Institute of Medicine). 1972. *Educating for the health team.* Washington, DC: National Academy Press. P. 4.

IOM. 2010. *Redesigning continuing education in the health professions.* Washington, DC: The National Academies Press.

Isaac, T., J. Zheng, and A. Jha. 2012. Use of UpToDate and outcomes in U.S. hospitals. *Journal of Hospital Medicine* 7(2):85–90.

Johnson, J. A., J. L. Bootman, W. E. Evans, R. A. Hudson, D. Knoell, L. Simmons, R. Straubinger, and S. M. Meyer. 2002. Pharmacogenomics: A scientific revolution in pharmaceutical sciences and pharmacy practice. Report of the 2001–2002 Academic Affairs Committee. *American Journal of Pharmaceutical Education* 66:12S–15S.

Kind, T., P. D. Patel, D. Lie, and K. C. Chretien. 2014. Twelve tips for using social media as a medical educator. *Medical Teacher* 36(4):284–290.

Korf, B. R., A. B. Berry, M. Limson, A. J. Marian, M. F. Murray, P. P. O'Rourke, E. R. Passamani, M. V. Relling, J. Tooker, G. J. Tsongalis, and L. L. Rodriguez. 2014. Framework for development of physician competencies in genomic medicine: Report of the Competencies Working Group of the Inter-Society Coordinating Committee for Physician Education in Genomics. *Genetics in Medicine* 16(11):804–809.

Kuo, G. M., J. D. Ma, K. C. Lee, J. R. Halpert, P. E. Bourne, T. G. Ganiats, and P. Taylor. 2011. Institutional profile: University of California San Diego Pharmacogenomics Education Program (PharmGenEd): Bridging the gap between science and practice. *Pharmacogenomics* 12(2):149–153.

Larsen, D. P., A. C. Butler, and H. L. Roediger, 3rd. 2008. Test-enhanced learning in medical education. *Medical Education* 42(10):959–966.

Logghe, H., J. Maa, and J. Schwartz. 2013. Twitter usage at Clinical Congress rises markedly over two years. *Bulletin of the American College of Surgery* 98(2):22–24.

Lowe, M. M., N. Bennett, and A. Aparicio. 2009. The role of audience characteristics and external factors in continuing medical education and physician change: Effectiveness of continuing medical education: American College of Chest Physicians evidence-based educational guidelines. *Chest* 135(3 Suppl):56s–61s.

Manolio, T. A., and M. F. Murray. 2014. The growing role of professional societies in educating clinicians in genomics. *Genetics in Medicine* 16(8):571–572.

Mansouri, M., and J. Lockyer. 2007. A meta-analysis of continuing medical education effectiveness. *Journal of Continuing Education in the Health Professions* 27(1):6–15.

Marinopoulos, S. S., T. Dorman, N. Ratanawongsa, L. M. Wilson, B. H. Ashar, J. L. Magaziner, R. G. Miller, P. A. Thomas, G. P. Prokopowicz, R. Qayyum, and E. B. Bass. 2007. Effectiveness of continuing medical education. *Evidence Report/Technology Assessment (Full Report)* 149:1–69.

Matta, R., C. Doiron, and M. J. Leveridge. 2014. The dramatic increase in social media in urology. *Journal of Urology* 192(2):494–498.

Mazmanian, P. E., and D. A. Davis. 2002. Continuing medical education and the physician as a learner: Guide to the evidence. *Journal of the American Medical Association* 288(9):1057–1060.

Mazmanian, P. E., D. A. Davis, and R. Galbraith. 2009. Continuing medical education effect on clinical outcomes: Effectiveness of continuing medical education: American College of Chest Physicians evidence-based educational guidelines. *Chest* 135(3 Suppl):49s–55s.

McCullough, K. B., C. M. Formea, K. D. Berg, J. A. Burzynski, J. L. Cunningham, N. N. Ou, M. I. Rudis, J. L. Stollings, and W. T. Nicholson. 2011. Assessment of the pharmacogenomics educational needs of pharmacists. *American Journal of Pharmaceutical Education* 75(3):51.

McInerney, J. D., E. Edelman, T. Nissen, K. Reed, and J. A. Scott. 2012. Preparing health professionals for individualized medicine. *Personalized Medicine* 9(5):529–537.

Moore, D. E., Jr., J. S. Green, and H. A. Gallis. 2009. Achieving desired results and improved outcomes: Integrating planning and assessment throughout learning activities. *Journal of Continuing Education in the Health Professions* 29(1):1–15.

Nasca, T. J., I. Philibert, T. Brigham, and T. C. Flynn. 2012. The next GME accreditation system—rationale and benefits. *New England Journal of Medicine* 366(11):1051–1056.

Neill, A., J. J. Cronin, D. Brannigan, R. O'Sullivan, and M. Cadogan. 2014. The impact of social media on a major international emergency medicine conference. *Emergency Medicine Journal* 31(5):401–404.

O'Donnell, P. H., K. Danahey, M. Jacobs, N. R. Wadhwa, S. Yuen, A. Bush, Y. Sacro, M. J. Sorrentino, M. Siegler, W. Harper, A. Warrick, S. Das, D. Saner, C. L. Corless, and M. J. Ratain. 2014. Adoption of a clinical pharmacogenomics implementation program during outpatient care—Initial results of the University of Chicago "1,200 Patients Project." *American Journal of Medical Genetics, Part C: Seminars in Medical Genetics* 166c(1):68–75.

Radecki, R. P., S. R. Rezaie, and M. Lin. 2014. *Global Emergency Medicine* Journal Club: Social media responses to the November 2013 *Annals of Emergency Medicine* Journal Club. *Annals of Emergency Medicine* 63(4):490–494.

Reiss, S. M. 2011. Integrating pharmacogenomics into pharmacy practice via medication therapy management. *Journal of the American Pharmacists Association* 51(6):e64–e74.

Roediger, H. L., 3rd, and A. C. Butler. 2011. The critical role of retrieval practice in long-term retention. *Trends in Cognitive Science* 15(1):20–27.

Rubenstein, L. V., and J. Pugh. 2006. Strategies for promoting organizational and practice change by advancing implementation research. *Journal of General Internal Medicine* 21(Suppl 2):S58–S64.

Rubenstein, L. V., B. S. Mittman, E. M. Yano, and C. D. Mulrow. 2000. From understanding health care provider behavior to improving health care: The QUERI framework for quality improvement. Quality Enhancement Research Initiative. *Medical Care* 38(6 Suppl 1):I129–I141.

Sansgiry, S. S., and A. S. Kulkarni. 2003. The human genome project: Assessing confidence in knowledge and training requirements for community pharmacists. *American Journal of Pharmaceutical Education* 67(2):1–10.

Shuldiner, A. R., K. Palmer, R. E. Pakyz, T. D. Alestock, K. A. Maloney, C. O'Neill, S. Bhatty, J. Schub, C. L. Overby, R. B. Horenstein, T. I. Pollin, M. D. Kelemen, A. L. Beitelshees, S. W. Robinson, M. G. Blitzer, P. F. McArdle, L. Brown, L. J. Jeng, R. Y. Zhao, N. Ambulos, and M. R. Vesely. 2014. Implementation of pharmacogenetics: The University of Maryland personalized anti-platelet pharmacogenetics program. *American Journal of Medical Genetics, Part C: Seminars in Medical Genetics* 166c(1):76–84.

Stetler, C. B., M. W. Legro, C. M. Wallace, C. Bowman, M. Guihan, H. Hagedorn, B. Kimmel, N. D. Sharp, and J. L. Smith. 2006. The role of formative evaluation in implementation research and the QUERI experience. *Journal of General Internal Medicine* 21(Suppl 2):S1–S8.

Toll, E. 2012. A piece of my mind. The cost of technology. *Journal of the American Medical Association* 307(23):2497–2498.

Yerkes, R. M., and J. D. Dodson. 1908. The relationship of strength of stimulus to rapidity of habit-formation. *Journal of Comparative Neurology and Psychology* 18(5):459–482.

A

Workshop Agenda

Improving Genetics Education in Graduate and Continuing Health Professional Education: A Workshop
August 18, 2014

The Keck Center of the National Academies, Room 100
500 Fifth Street, NW
Washington, DC 20001

MEETING OBJECTIVES

Examine context for the challenges involved in educating health care providers in genetics

- Review promising approaches for providing genetics education in various settings.
- Identify opportunities and next steps for improving genetics education for health professionals.

AGENDA

8:30–8:35 a.m. **Welcoming remarks**

Sharon F. Terry, *Roundtable Co-Chair*
President and Chief Executive Officer
Genetic Alliance

Geoffrey Ginsburg, *Roundtable Co-Chair*
Executive Director, Center for Personalized Medicine, Duke Medicine; Professor of Medicine and Pathology, Duke University Medical Center

8:35–8:50 **Charge to workshop speakers and participants**

Bruce Blumberg, *Workshop Co-Chair*
Institutional Director of Graduate Medical Education, Northern California
Kaiser Permanente, The Permanente Medical Group

Joan A. Scott, *Workshop Co-Chair*
Chief, Genetic Services Branch
Division of Services for Children with Special Health Needs
Maternal and Child Health Bureau
Health Resources and Services Administration

8:50–9:10 **Mistakes and myths in graduate and continuing professional education: What we might do differently and what will be effective**

David Davis
Senior Director, Continuing Education and Performance Improvement
Association of American Medical Colleges

SESSION I: EDUCATIONAL APPROACHES

Session objectives:

- To identify educational processes and innovative models that could be applied to improve genetics education of health professionals in various settings.
- To examine opportunities for changing provider behavior through genetics education.
- To identify barriers to implementing educational approaches in genetics.

Moderator: David Davis, AAMC

9:10–9:55	**Challenges in reaching community practitioners** Perry A. Pugno Vice President for Education American Academy of Family Physicians **Just-in-time approaches to education** Benjamin Raby Section Editor–Genetics UpToDate Associate Professor of Medicine Channing Division of Network Medicine and the Division of Pulmonary and Critical Care Medicine, Director, Brigham and Women's Hospital Pulmonary Genetics Center Harvard Medical School **Innovative models of education: Using technology appropriately in medical education** Alexander M. Djuricich Associate Dean for Continuing Medical Education Program Director, Medicine–Pediatrics Residency Associate Professor for Clinical Pediatrics and Clinical Medicine Indiana University School of Medicine
9:55–10:10	**BREAK**
10:10–10:55	**Interprofessional education in genetics** Diane C. Seibert Professor Chair and Director, Family Nurse Practitioner Program Uniformed Services University of the Health Sciences

Marketing or education?: Identifying and avoiding conflicts of interest in educational materials

Jean Silver-Isenstadt
Executive Director
National Physicians Alliance

Principles of evidence in designing educational programs

Maren T. Scheuner
Chief, Medical Genetics
Veterans Affairs Greater Los Angeles Healthcare System

10:55–11:55 **Discussion with speakers and attendees**

Perry A. Pugno
American Academy of Family Physicians

Benjamin Raby
Harvard Medical School and UpToDate

Alexander M. Djuricich
Indiana University School of Medicine

Diane C. Seibert
Uniformed Services University of the Health Sciences

Jean Silver-Isenstadt
National Physicians Alliance

Maren T. Scheuner
Veterans Affairs Greater Los Angeles Healthcare System

11:55 a.m.–12:55 p.m. **WORKING LUNCH**

SESSION II: GRADUATE HEALTH PROFESSIONAL EDUCATION/POST-GRADUATE TRAINING

Session objective:

- To discuss pragmatic approaches to the application of educational processes and principles for graduate/post-graduate genetics education.

Moderator: Bruce Blumberg, Kaiser Permanente

12:55–1:40 Kevin B. Weiss
Senior Vice President for Institutional Accreditation Services
Accreditation Council for Graduate Medical Education

Grace M. Kuo
Professor of Clinical Pharmacy and Associate Dean for Academic Clinical Affairs
Adjunct Professor of Family and Preventive Medicine
University of California, San Diego

Rebecca S. Lipner
Senior Vice President, Evaluation, Research, and Development
American Board of Internal Medicine

1:40–2:30 **Discussion with speakers and attendees**

Samantha Bazan
Uniformed Services University of the Health Sciences
Graduate School of Nursing

Kevin B. Weiss
Accreditation Council for Graduate Medical Education

Grace M. Kuo
University of California, San Diego

Rebecca S. Lipner
American Board of Internal Medicine

2:30–2:45 **BREAK**

SESSION III: CONTINUING PROFESSIONAL DEVELOPMENT

Session objective:

- To discuss pragmatic approaches to the application of educational processes and principles for continuing professional genetics education.

Moderator: Sam Johnson, Kaiser Permanente

2:45–3:30 Ann Karty
Chair, Continuing Professional Development Directors Group
Council of Medical Specialty Societies

Murray Kopelow
President and Chief Executive Officer
Accreditation Council for Continuing Medical Education

Constance Goldgar
Associate Director
University of Utah Physician Assistant Program

3:30–4:15 **Discussion with speakers and attendees**

Ann Karty
Council of Medical Specialty Societies

Murray Kopelow
Accreditation Council for Continuing Medical Education

Constance Goldgar
University of Utah Physician Assistant Program

Benjamin Raby
Harvard Medical School and UpToDate

SESSION IV: JOINT ACADEMY/PROFESSIONAL SOCIETY APPROACH TO GENETICS EDUCATION

4:15–4:30 Michael Murray
Co-chair
Inter-Society Coordinating Committee for Practitioner Education in Genomics

SESSION V: NEXT STEPS TO ACHIEVE EFFECTIVE GENETICS EDUCATION FOR HEALTH PROFESSIONALS

Moderator: Michael Murray, Inter-Society Coordinating Committee for Practitioner Education in Genomics

4:30–5:20 **What is the needle we need to move and how do we move it?**

Murray Kopelow
Accreditation Council for Continuing Medical Education

Kevin B. Weiss
Accreditation Council for Graduate Medical Education

Grace M. Kuo
University of California, San Diego

Diane C. Seibert
Uniformed Services University of the Health Sciences

Constance Goldgar
University of Utah Physician Assistant Program

5:20–5:35 **Summary and concluding remarks**

Bruce Blumberg, *Workshop Co-Chair*
Institutional Director of Graduate Medical Education, Northern California
Kaiser Permanente, The Permanente Medical Group

Joan A. Scott, *Workshop Co-Chair*
Chief, Genetic Services Branch
Division of Services for Children with Special Health Needs
Maternal and Child Health Bureau
Health Resources and Services Administration

5:35 **ADJOURN**

B

Speaker Biographical Sketches

Samantha Bazan, M.S., earned her bachelor of science in nursing from George Mason University in 2004. She received a master of science with a concentration in disaster management from Trident University in 2012. In May 2004, she commissioned into the Army Nurse Corps. She started her nursing career at Fort Sill, Oklahoma, as a medical–surgical nurse and quickly deployed to Baghdad, Iraq, with the 10th Combat Support Hospital working in the intensive care unit. She has also worked in the emergency room and specialized as an army public health nurse in 2008. Samantha Bazan is still on active duty. In 2013 she was accepted into the Doctor of Nursing Practice for Family Nurse Practitioner program at the Uniformed Services University of the Health Sciences. Her awards include Meritorious Service Medal, the Army Commendation Medal with two Oak Leaf Clusters, the Army Achievement Medal, the Overseas Service Ribbon, Iraq Campaign Medal, Global War on Terrorism Service Medal, National Defense Service Ribbon, and the Army Service Ribbon. She served with one unit that received the Meritorious Unit Commendation.

Bruce D. Blumberg, M.D., is the director of graduate medical education (the resident physician training programs) for Northern California Kaiser Permanente. He currently maintains a clinical practice in medical genetics at Kaiser Permanente Oakland. He is a clinical professor of pediatrics at the University of California, San Francisco, and an adjunct clinical professor of pediatrics at Stanford University School of Medicine. His clinical interests within genetics are broad, and he has a subspecialty interest in inherited disorders of skeletal and connective tissue development. His research interest is in the area of the psychosocial and emotional aspects of prenatal diagnosis. Dr. Blumberg holds a medical degree from Yale University School of Medicine and completed his residency in pediatrics at Stanford University Hospital and the University of California, Los Angeles, Center for the Health Sciences as well as a fellowship in medical genetics at Harbor–UCLA Medical Center. He also received a B.A. from Dartmouth College.

David A. Davis, M.D., FCFP, is the senior director of continuing education and performance improvement at the Association of American Medical Colleges (AAMC). Prior to this role, Dr. Davis was a family physician in Ontario, Canada, for nearly 40 years. For much of that time, he was active in continuing medical education (CME) as chairman of an all-staff interprofessional continuing education program at a community hospital; director of CME and subsequently chair of continuing education at McMaster University's Faculty of Health Sciences; associate dean of continuing education and founding director of the Knowledge Translation (Implementation Science) Program in the Faculty of Medicine, University of Toronto; and chairman of Ontario's Guidelines Advisory Committee. This last role allowed Dr. Davis to explore and test models of clinical practice guideline development, adaptation, and implementation of best evidence on a province-wide basis. Dr. Davis has also developed an innovative comprehensive competency assessment program for the provincial licensing body, and he helped create a center for faculty development and a mini-med school at the University of Toronto. Emphasizing the evaluation of educational activities using a rigorous outcomes-testing approach, he has acted as principal investigator, co-principal investigator, or investigator on grants totaling several million dollars. This emphasis has seen the publication of 125 peer-reviewed papers in addition to dozens of abstracts, book chapters, 2 major books on CME practices, and presentations on four continents. His (and colleagues') 1995 *Journal of the American Medical Association* systematic review of the effect of CME interventions is widely cited as a seminal study in this field. Finally, Dr. Davis has been chair or president of national or provincial Canadian organizations, two North American organizations (the Alliance for Continuing Medical Education and the Society for Academic Continuing Medical Education) and the Guidelines International Network, a global organization dedicated to the development and implementation of evidence-based clinical practice guidelines. His current role in AAMC permits him the opportunity to work with individuals, associations, groups, and academic medical centers to create scholarly, integrated models of effective, performance-based continuing education.

Alexander M. Djuricich, M.D., the associate dean for continuing medical education at the Indiana University School of Medicine (IUSM), grew up in Chicago. After undergraduate work at Northwestern University, he completed medical school at Loyola University Stritch School of Medicine in Chicago in 1998, then came to IUSM for his residency in combined internal medicine–pediatrics ("med–peds," for short) from 1994 to 1998. He worked as a primary care med–peds physician from 1998 to 2001, also completing the Michigan State Primary Care Faculty Development Fellowship Program in 2001. He returned to IUSM in 2001, with a growing interest in resident education over the next 10 years, first as associate program director, then as program director of the Medicine–Pediatrics Residency Program. He was the medical director for quality improvement at Riley Hospital for Children from 2006 to 2011. He served as the president of the Medicine–Pediatrics Program Directors' Association from 2010 to 2011. His areas of interest include quality improvement and patient safety, emerging technology in medicine and medical education, health care provider involvement in social media, and medical education for residents and faculty.

Geoffrey Ginsburg, M.D., Ph.D., is the founding director for the Center for Applied Genomics in the Duke University Medical Center and the founding executive director of the Center for Personalized and Precision Medicine in the Duke University Health System. He is a professor of medicine, pathology, and biomedical engineering at Duke University. He is an internationally recognized expert in genomics and personalized medicine with funding from the National Institutes of Health, Department of Defense, Air Force, Defense Advanced Research Projects Agency, the Gates Foundation, and industry. Prior to Duke he was at Millennium Pharmaceuticals Inc., where he was vice president of molecular and personalized medicine and responsible for developing pharmacogenomic and biomarker strategies for therapeutics. He serves as an expert panel member for Genome Canada, as a member of the board of external experts for the National Heart, Lung, and Blood Institute, as co-chair of the IOM's Roundtable on Translating Genomic-Based Research for Health, as a member of the advisory council for the National Center for Advancing Translational Sciences, as co-chair of the Cures Acceleration Network, as an advisor to the Pharmacogenetics Research Network, and as a member of the World Economic Forum's Global Agenda Council on the Future of the Health Sector.

Constance Goldgar, M.S., PA-C, is an associate professor and associate director at the University of Utah Physician Assistant Program in the Department of Family and Preventive Medicine. Before coming to physician assistant education 15 years ago, Ms. Goldgar worked in genetic epidemiology research for 8 years at the University of Utah. Her areas of expertise in teaching are genetics and evidence-based medicine. She helped author an interactive educational website sponsored by the National Coalition for Health Professional Education in Genetics for physician assistants, through which, in part, she received the Michael J. Scotti, Jr. Award. She is also an editorial and advisory member for the National Institutes of Health–funded interprofessional Genetics Genomics Competency Center for Education.

Samuel G. Johnson, Pharm.D., BCPS, earned his B.S. in biology from Truman State University in 1998, followed by earning his Pharm.D. from the University of Missouri–Kansas City in 2003. Outside of academics, Dr. Johnson was very active in several student organizations: the American Pharmaceutical Association–Academy of Students of Pharmacy, the Student Society for Health Systems Pharmacists, and Kappa Psi, among others. Dr. Johnson embarked on his professional career at Kaiser Permanente Colorado (KPCO) in 2004 as a clinical pharmacy specialist. For the next 2.5 years he maintained an active clinical and teaching practice within the Clinical Pharmacy Anticoagulation Service, until accepting a position in late 2006 as clinical pharmacy specialist in cardiology. In 2011 he left his post in cardiology for a newly created position as a clinical pharmacy specialist in applied pharmacogenomics, and he is currently responsible for leading the clinical implementation efforts for the application of pharmacogenomics within KPCO. In addition to his daily work responsibilities, he is actively engaged in teaching at two local schools of pharmacy and in research efforts for the Clinical Pharmacy Research Team. In 2010 he was awarded the Kaiser Permanente Colorado Summit Award for outstanding individual achievement. In 2009 he was awarded the "RxCellence" Award in recognition of outstanding contributions to the Pharmacy Department. Dr. Johnson is a past secretary/treasurer for the Cardiology Practice and Research Network of the American College of Clinical Pharmacy and has served as StuNet Liaison since 2008. He also currently serves as a vice chair for the Colorado Medicaid Drug Utilization Review Board as well as the community practitioner representative for the University of Colorado Skaggs School of Pharmacy and Pharmaceutical Sciences Non-Traditional Pharm.D. Program Committee.

Ann Karty, M.D., FAAFP, joined the American Academy of Family Physicians (AAFP) in 2009 and leads AAFP's continuing medical education (CME) efforts by representing AAFP in its external relationships with other accrediting and educational organizations. Along with oversight of AAFP CME activity content, Dr. Karty works with teams creating innovative educational formats, CME planning at the annual AAFP assembly, and the AAFP credit system. She is currently involved in the risk evaluation mitigation strategies team at AAFP and has represented the academy at national organizations, including FDA and the Council of Medical Specialty Societies, where she also currently serves as the chair of the CME director component group. Just prior to joining the AAFP staff, Dr. Karty served as an associate professor in the Department of Family Medicine at the Kansas City University of Medicine and Biosciences College of Osteopathic Medicine (KCUMB) while she maintained her own private practice. Dr. Karty also served as the associate program director for the Medical Center of Independence–KCUMB family medicine residency program and has had multiple academic appointments at the state level. Dr. Karty has been an active member of AAFP since 1989. Throughout her career Dr. Karty has presented and moderated numerous CME sessions at the local and national level. Internationally, Dr. Karty serves as a co-chair of the Hadassah Physicians' Council and has served many years on the steering committee for the Hadassah CME committee, which coordinated its first CME meeting hosted by Israeli physicians in 2008. Since that time she has moderated several of the CME sessions in Israel, and Dr. Karty is currently serving her fourth term as co-chair for the November 2014 international meeting. Dr. Karty is board certified in family medicine and is a fellow of AAFP. She holds medical licenses in Illinois, Kansas, Missouri, and South Carolina. She received a bachelor of arts degree in biology and a medical degree from the University of Missouri–Kansas City School of Medicine and is currently pursuing a master of business administration from the University of Kansas School of Business.

Murray Kopelow, M.D., M.S. (Comm), FRCPC, is the president and chief executive officer of the Accreditation Council for Continuing Medical Education (ACCME), where he leads ACCME's efforts to identify, develop, and promote national standards for quality continuing medical education (CME) that improves physician competence and performance and contributes to improving medical care for patients and their communities. Dr. Kopelow is responsible for the implementation of the ACCME system for accrediting U.S. institutions that offer CME and for the ACCME system of recognizing state and territory medical societies as accreditors for intrastate CME providers. Dr. Kopelow has overseen the evolution of the accreditation system, including the 2004 update of the ACCME's Standards for Commercial Support: Standards to Ensure Independence and the introduction of the 2006 Accreditation Criteria, which position accredited CME as a Bridge to Quality™. Dr. Kopelow has advised CME accreditation systems around the world and collaborated on the creation of a substantial equivalency recognition process for CME systems outside the United States. From June through December 2009, Dr. Kopelow served as a special advisor to the Office of Demand Reduction within the Office of National Drug Control Policy of the Executive Office of the President of the United States. From 2010 through the present, he has worked with the FDA, health care education accreditors, and industry to facilitate the role of accredited CME in supporting FDA's risk evaluation and mitigation strategy for opioid medications. A native of Canada, Dr. Kopelow holds a medical degree from the Faculty of Medicine, University of Manitoba, and a master's of science in communications systems from the Department of Communications Studies at Northwestern University. He is a fellow of the Royal College of Physicians and Surgeons of Canada and a successful participant in that organization's maintenance of certification program.

Grace M. Kuo, Pharm.D., M.P.H., Ph.D., FCCP, received her bachelor of science degree in psychobiology from the University of California, Los Angeles, followed by her bachelor of science degree in pharmacy from the Massachusetts College of Pharmacy. She then completed her doctor of pharmacy degree from Oregon State University, Oregon Health Sciences University. Her postdoctoral training was at the W.G. Magnuson Clinical Center at the National Institutes of Health in Maryland. She subsequently earned both her master of public health degree and her doctor of philosophy degree in public health from the University of Texas Health Science Center at Houston School of Public Health. Currently she is a professor of clinical pharmacy, adjunct professor of family and preventive medicine, and associate dean for academic clinical affairs at the University of California, San Diego, Skaggs School of Pharmacy and Pharmaceutical Sciences. Dr. Kuo's clinical expertise is primary care practice. Being a pharmacist health services researcher, Dr. Kuo devotes her efforts in practice-based research that focuses on medication safety and medication therapy management. Working with experts and national leaders, Dr. Kuo and her team have developed and administered the national pharmacogenomics education program (PharmGenEd). Using a peer-reviewed shared curriculum, the PharmGenEd team disseminates emerging scientific information about pharmacogenomics applicable to patient care. To date, PharmGenEd has been used by more than 3,000 health care professionals and faculty from 86 health professional schools.

Rebecca S. Lipner, Ph.D., is the senior vice president of evaluation, research, and development for the American Board of Internal Medicine (ABIM), where she oversees the scoring, statistical analysis, standard setting, equating, security, and evaluation of measurement properties for ABIM assessment products. She also oversees quantitative research analysis ranging from internal medicine workforce trends to health outcomes research. Dr. Lipner has expertise in performance measurement, working with clinical and patient experience/satisfaction survey data on the scientific acceptability of a measure's properties, including reliability, validity, risk adjustment, usability, and sampling strategies. She has developed a unique weighted average composite score and standard-setting methodology for assessing physician performance, which has been patented. Dr. Lipner also has expertise in the field of measurement in general, including testing of physician's problem-solving skills through clinical vignettes as well as procedural skills through high-fidelity simulation. Dr. Lipner is a frequent speaker on these subjects and is widely published in professional journals, including the *Journal of the American Medical Association*, *Annals of Internal Medicine*, *Academic Medicine*, *Applied Measurement in Education*, and the *Journal of Educational Measurement*. Prior to joining ABIM, Dr. Lipner held a variety of teaching and faculty positions at Drexel University, St. Joseph's University, and the University of Pittsburgh, where she taught undergraduate and graduate courses in statistics, tests and measurement, experimental design, systems analysis and design, and expert systems.

Michael F. Murray, M.D., is the co-chair of the Inter-Society Coordinating Committee for Practitioner Education in Genomics (www.genome.gov/27554614), which was launched in 2013, through the leadership of the National Human Genome Research Institute, with the goal of bringing together professional organizations within clinical medicine to improve genomic literacy. Dr. Murray is boarded in internal medicine and medical genetics and he joined Geisinger Health System in Pennsylvania as the director of clinical genomics in 2013 after serving as the clinical chief of genetics at Brigham and Women's Hospital in Boston for 9 years. While in Boston he launched and directed the Harvard Medical School continuing medical education course "The Genetic and Genomic Basis of Adult Medicine: What the Primary Care Provider Needs to Know." He also served as the first program director of the combined residency in internal medicine and medical genetics, which was part of the Harvard Genetics Training Program. He is the lead editor of the genomics textbook for practicing clinicians, *Clinical Genomics: Practical Applications for Adult Patient Care* (McGraw-Hill, 2013). His patient care responsibilities have included running the Adult Genetics Clinic at Brigham and Women's Hospital and initiating a personalized genomic consult service there. At Geisinger he is leading the clinical return of results program for 100,000 participants who will undergo genomic sequencing as part of the health system's biobank program (MyCode).

Perry A. Pugno, M.D., M.P.H., FAAFP, FACEP, FACPE, is a 1970 graduate of the University of California, Riverside, and a 1974 graduate of the University of California, Davis, School of Medicine. Dr. Pugno completed his University of California, Los Angeles–affiliated family medicine residency at Ventura General Hospital. Following a tour of duty with the National Health Service Corps in Barstow, California, he entered the sphere of graduate medical education as a residency director, and he has accumulated more than 20 years of experience in that role. He has worked in programs from California to Connecticut, including public, private, and university-sponsored settings. He is board certified in both family medicine and emergency medicine, and he has added experience as the director of a trauma center, a hospital chief medical officer, a public health officer, and a medical director of a health plan. His M.P.H. from Loma Linda University is in multidisciplinary educational administration, and he is a fellow of the American College of Physician Executives. He has served as the president of the Association of Family Medicine Residency Directors, president of the University of California Medical Alumni Association, and chair of the Residency Review Committee for Family Medicine at the Accreditation Council for Graduate Medical Education, and he is the founding chair of the National Institute for Program Director Development. His previous experience in corporate physician leadership and managed care was as the vice president for graduate medical education and medical affairs with Mercy Healthcare Sacramento, a division of Catholic Healthcare West (now Dignity Health). Dr. Pugno retired in 2014 from the position of vice president for education for the American Academy of Family Physicians (AAFP). In that position he was responsible for AAFP initiatives related to medical school, graduate, and continuing medical education, including supervision of the Residency Program Solutions consulting panel and providing staff direction for academy workforce policy and graduate medical education advocacy.

Benjamin Raby, M.D., is a pulmonologist and genetic epidemiologist with expertise in the genetics and genomics of asthma. He is an associate professor of medicine at the Brigham and Women's Hospital (BWH) and Harvard Medical School. He is the founder and director of the BWH Pulmonary Genetics Center and the principal investigator of multiple National Heart, Lung, and Blood Institute–sponsored grants focused on genomic approaches in asthma and other lung diseases, and he has published more than 120 original science manuscripts in these fields. He is section editor for genetics at UpToDate, Inc., and he is an editorial board member of several subspecialty journals, including the *American Journal of Respiratory and Critical Care Medicine* and *Allergy*.

Maren T. Scheuner, M.D., M.P.H., is an internist and medical geneticist specializing in the field of common disease genetics with more than 20 years of experience providing clinical genetic services. She divides her time between her clinical practice of adult genetics and health services and implementation research. She began her research career in 2005 and she has had continuous funding from the National Institutes of Health (the National Human Genome Research Institute, National Cancer Institute, and National Heart, Lung, and Blood Institute), the Centers for Disease Control and Prevention, and the Department of Veterans Affairs ever since. She has authored more than 50 publications and is a nationally recognized expert on adult genetics and health services genomics. Her areas of interest include genetic risk assessment for chronic diseases of adulthood; health services and policy research in genetic/genomic medicine; assessment of genetic/genomic technologies; health information technology and genomics; development and evaluation of family history tools for public health and preventive medicine practice; implementation, surveillance and outcomes research in medical genetics/genomics; and professional education in genetics/genomics.

Joan A. Scott, M.S., C.G.C., is the chief of the Genetic Services Branch in the Division of Children with Special Health Needs of the Maternal and Child Health Bureau at the Health Resources and Services Administration (HRSA). She is a certified genetic counselor with more than 35 years' experience in clinical genetics, genetics education, laboratory medicine, the biotechnology industry, and the ethical, legal, social, and policy implications of advances in genomics. Ms. Scott's career has focused on the application of genomic discoveries to health care. Prior to coming to HRSA, she was the executive director of the National Coalition for Health Professional Education in Genetics (NCHPEG), where she led a national effort to promote health professional education and access to information about advances in human genetics, and she was a research scientist in the Berman Institute of Bioethics at Johns Hopkins University, where she studied public and stakeholder attitudes about genomics. Prior to joining NCHPEG, Ms. Scott was the director of the Genetics and Public Policy Center at Johns Hopkins University, which was established to fill an important niche in the science policy landscape. There she led the center's efforts to address policy issues related to advances in genetics, genetic testing quality and oversight, and public engagement in genetic research. Prior to coming to the center in 2002, Ms. Scott was a director in GeneLogic, Inc., overseeing the operations of a large biorespository for use in genomic discovery. She also served as general manager and director of genetic services at the clinical diagnostic lab OncorMed from 1994 to 1998. Clinically, she has practiced in a variety of academic, outreach, and private practice settings, including pediatric, adult, and reproductive genetic clinics. Ms. Scott is a past president of the National Society of Genetic Counselors and a founding member of the American Board of Genetic Counseling. She has served on numerous national committees and work groups, including the Evaluation of Genomic Applications in Practice and Prevention Working Group; the Secretary's Advisory Committee on Genetics, Health, and Society Task Force on DTC Genetic Testing; the Maryland Insurance Administration Workgroup on Genetic Testing; the National Cancer Institute's CaHUB

Advisory Committee; and the Genetic Alliance Biobank Advisory Board. Ms. Scott holds an M.S. (Human Genetics Program) from Sarah Lawrence College and a B.A. in anthropology and zoology from Kent State University. She has been certified by the American Board of Medical Genetics with a subspecialty in genetic counseling, and she was recertified by the American Board of Genetic Counseling in 2006.

Diane C. Seibert, Ph.D., ARNP, FAAN, FAANP, is a professor in and the chair of the family nurse practitioner program at the Uniformed Services University of the Health Sciences in Bethesda, Maryland. She is certified as both a women's health and an adult nurse practitioner, and she maintains an active clinical practice at the National Naval Medical Center in Bethesda, Maryland. Dr. Seibert has published and presented to a variety of audiences on women's health and genetics. She helped establish national and international provider competencies, developed new curriculums to ensure quality and consistent genomic care, and is involved in several national task forces and committees working toward improving the genetics competency of the nursing workforce across all practice settings. She has played a key role in the development of international genomic research priorities and a pioneering practice change model in collaboration with the U.S. Genetic/Genomic Nursing Competency Initiative. Dr. Seibert received her B.S.N. from Kent State University, her master's degree from the University of Maryland at Baltimore, and her Ph.D. from the University of Maryland, College Park.

Jean Silver-Isenstadt, M.D., Ph.D., serves as the executive director of the National Physicians Alliance and was actively involved in the organization's founding. She holds a doctorate in the history and sociology of medicine from the University of Pennsylvania, a medical degree from the University of Maryland, and a master's degree in nonfiction and science writing from Johns Hopkins University. Her doctoral work focused on 19th-century American health reform. She is the author of *Shameless: The Visionary Life of Mary Gove Nichols* (Johns Hopkins University Press, 2002), a biography of the infamous and influential health advocate and social reformer best known for her leadership of the water-cure movement and for her scandalous public lectures to women on anatomy and physiology.

Sharon Terry, M.A., is the president and chief executive officer of the Genetic Alliance, a network of more than 10,000 organizations, 1,200 of which are disease advocacy organizations. Genetic Alliance improves health through the authentic engagement of communities and individuals. It develops innovative solutions through novel partnerships, connecting consumers to smart services. She is the founding chief executive officer of PXE International, a research advocacy organization for the genetic condition pseudoxanthoma elasticum (PXE). As co-discoverer of the gene associated with PXE, she holds the patent for ABCC6 and has assigned her rights to the foundation. She developed a diagnostic test and is conducting clinical trials. Ms. Terry is also a co-founder of the Genetic Alliance Registry and Biobank. She is the author of more than 90 peer-reviewed articles. In her focus at the forefront of consumer participation in genetics research, services, and policy, she serves in a leadership role on many of the major international and national organizations, including the IOM Board on Health Sciences Policy, the National Coalition for Health Professional Education in Genetics board, and the International Rare Disease Research Consortium Interim Executive Committee, and she is co-chair of the IOM Roundtable on Translating Genomic-Based Research for Health. She is on the editorial boards of several journals. She was instrumental in the passage of the Genetic Information Nondiscrimination Act. In 2005 she received an honorary doctorate from Iona College for her work in community engagement; in 2007, she was awarded the first Patient Service Award from the University of North Carolina Institute for Pharmacogenomics and Individualized Therapy; in 2009, she received the Research!America Distinguished Organization Advocacy Award; and in 2011, she won the Clinical Research Forum and Foundation's Annual Award for Leadership in Public Advocacy. She is an Ashoka Fellow.

Kevin B. Weiss, M.D., M.P.H., has devoted his medical career to issues of health care quality, equity and access to care, and training physicians and other health care providers in health care improvement. As the senior vice president for institutional accreditation at the Accreditation Council for Graduate Medical Education (ACGME), he is responsible for the new Clinical Learning Environment Review (CLER) program. He also co-chairs the CLER evaluation committee and oversees the ACGME's Institutional Review Committee's accreditation activities. Prior to coming to ACGME, Dr. Weiss served as president and chief executive officer of the American Board of Medical Specialties (ABMS) from 2007 to 2012. While at ABMS he broadened public involvement in the board's activities; implemented both its ethics and professionalism and health and public policy programs; established alignment with maintenance of licensure; and, as part of health care reform, aligned maintenance of certification with the Medicare Physician Quality Reporting Initiative and established ABMS–International. He has served various roles on committees for the National Quality Forum, the National Committee for Quality Assurance, and the American Medical Association's Physicians Consortium for Performance Improvement. He has served as a member of the American College of Physicians' board of regents and chaired its committees for clinical guidelines and performance measurement. Dr. Weiss currently serves on the board of directors for the Educational Commission for Foreign Medical Graduates and has served on committees for the IOM, including those which developed the reports *Crossing the Quality Chasm* and *Identifying Priority Areas for Quality Improvement*. Over the years Dr. Weiss has conducted federally funded U.S. and international epidemiological and health services research projects related to guideline implementation, chronic care management, outcomes measurement, quality improvement, and health care equity and has published more than 200 articles, reviews, books, book chapters, and monographs. In 2005 Dr. Weiss established the first U.S. graduate-level master's and Ph.D. degree programs in patient safety and health care quality at Northwestern University. Dr. Weiss is certified in internal medicine by the American Board of Internal Medicine. He also maintains a role as a professor of clinical medicine in the Division of General Medicine and in the Center for Healthcare Studies in the Feinberg School of Medicine at Northwestern University.

C

Statement of Task

An ad hoc committee will organize and conduct a public workshop to examine pragmatic approaches to improving genetics education in both graduate and continuing health professional education. The workshop goals will be to examine different educational approaches, analyze challenges to success, and explore potential solutions for improving genetics education for graduate and health care professionals. A diverse stakeholder group will be invited to present their perspectives. The committee will develop the workshop agenda, select speakers and discussants, and moderate the discussions. An individually authored summary of the presentations and discussions at the workshop will be prepared by a designated rapporteur in accordance with institutional guidelines.

D

Registered Attendees

Yawo Akrodou
Student

Naomi Aronson
Blue Cross and Blue Shield Association

Laurie Badzek
American Nurses Association

Samanta Bazan
Uniformed Services University of the Health Sciences

Laura Beamer
Northern Illinois University School of Nursing and Health Studies

Judith Benkendorf
American College of Medical Genetics and Genomics

Rebecca Blanchard
Merck and Co., Inc.

Miriam Blitzer
American Board of Medical Genetics and Genomics

Bruce D. Blumberg
Kaiser Permanente

Kathleen Calzone
National Cancer Institute, Center for Cancer Research, Genetics Branch

Colleen Campbell
Iowa Institute of Human Genetics

Ann Cashion
National Institute of Nursing Research
National Institutes of Health

C. Thomas Caskey
Baylor College of Medicine

Joowon Choi
Genetic Alliance

Elizabeth Cohn
Adelphi University
Columbia University

Patricia Cuff
Institute of Medicine

David Davis
Association of American Medical Colleges

Alexander Djuricich
Indiana University School of Medicine

Michael Dougherty
American Society of Human Genetics

Emily Edelman
The Jackson Laboratory

Julie Eggert
Clemson University

Greg Feero
Journal of the American Medical Association

Kelli Fee-Schroeder
Mayo Clinic

David Flannery
American College of Medical Genetics and Genomics

Geoffrey Ginsburg
Duke University

Constance Goldgar
University of Utah Physician Assistant Program

Susan Hahn
American Board of Genetic Counseling

Jennifer Hall
International Society for Cardiovascular Translational Research

Alyson Hanish
National Institute of Nursing Research/University of Iowa

Tiffany Harrington
Personalized Medicine Coalition

Elizabeth Hassen
Clemson University

Gillian Hooker
Next GxDx

Jennifer Hoskovec
National Society for Genetic Counselors

Haydee Jaramillo
Andromeda

Jean Jenkins
National Human Genome Research Institute

Samuel Johnson
Kaiser Permanente Colorado

Francis Kalush
U.S. Food and Drug Administration

Ann Karty
American Academy of Family Physicians

Jill Kaufman
College of American Pathologists

Esther Kim
23andMe

Murray Kopelow
Accreditation Council for Continuing Medical Education

Grace Kuo
University of California, San Diego, Skaggs School of Pharmacy and Pharmaceutical Sciences

Gabriela Lavezzari
Pharmaceutical Research and Manufacturers of America

Debra Leonard
Fletcher Allen Health Care
University of Vermont College of Medicine

Rebecca Lipner
American Board of Internal Medicine

Robert McCormack
Janssen Oncology

Kathleen McCormick
SciMind, LLC

Michael Miller
HealthPolCom

Michael Murray
Geisinger Health System

Carmen Paniagua
National Organization of Nurse Practitioner Faculties

Erin Payne
Northrop Grumman

Robert Plenge
Merck Research Labs

Victoria Pratt
Association for Molecular Pathology

Ronald Przygodzki
Department of Veterans Affairs

Perry Pugno
American Academy of Family Physicians

Benjamin Raby
Harvard Medical School

Jane Radford
American College of Medical Genetics

Nalini T. Raghavachari
National Insitute on Aging
National Institutes of Health

Kate Reed
The Jackson Laboratory

Allen Roses
Duke University

Beth Ruedi
Genetics Society of America

Lyndzie Sardenga
Genetic Alliance

Nadeem Sarwar
Eisai, Inc.

Maren Scheuner
Veterans Affairs Greater Los Angeles Healthcare System

Joan Scott
Health Resources and Services Administration

Diane Seibert
Uniformed Services University of the Health Sciences

Geetha Senthil
National Institute of Mental Health

Lisa Shah
University of Iowa

Jean Silver-Isenstadt
National Physicians Alliance

Jared Stevenson
Genetic Alliance

Katie Johansen Taber
American Medical Association

Sharon Terry
Genetic Alliance

Carolyn (Caer) Vitek
Mayo Clinic

Sam Wang
Genetic Alliance

Michael Watson
American College of Medical Genetics and Genomics

Jennifer Weisman
Strategic Analysis, Inc.

Kevin Weiss
Accreditation Council for Graduate Medical Education

Catherine Wicklund
Northwestern University

Janet Williams
University of Iowa
American Academy of Nursing

Michelle Wright
Virginia Commonwealth University

Janey Youngblom
California State University, Stanislaus